ENJOYING

RUM

Inspiring | Educating | Creating | Entertaining

Brimming with creative inspiration, how-to projects, and useful information to enrich your everyday life, Quarto Knows is a favorite destination for those pursuing their interests and passions. Visit our site and dig deeper with our books into your area of interest: Quarto Creates, Quarto Cooks, Quarto Homes, Quarto Lives, Quarto Drives, Quarto Explores, Quarto Gifts, or Quarto Kids.

25 24 23 22 21   1 2 3 4 5

ISBN: 978-0-7603-6928-9

Digital edition published in 2021
eISBN: 978-0-7603-6929-6

Library of Congress Control Number: 2021939543

Design and Page Layout: Ashley Prine, Tandem Books
Cover Frame: Vasya Kobelev/Shutterstock
Cover Illustration: NataLima/Shutterstock
Interior Illustrations and Photography: See page 176

Printed in China

ENJOYING

RUM

A TASTING GUIDE AND JOURNAL

FRANK FLANNERY

VOYAGEUR
PRESS

CONTENTS

TASTING NOTES

INTRODUCTION

Welcome to *Enjoying Rum: A Tasting Guide and Journal.* This book is your map to a liquor that has traveled the world, been used as a currency, fueled high-seas adventures and rebellions, cured what ailed you if it didn't kill you, and is at long last getting its due as a respectable libation able to stand on its own peg leg instead of being dismissed as a mixer. These pages explore the history of this sugarcane spirit, a story that is as deep and complex as rum's finest bottles, and sometimes just as dark. Once you find out where rum came from, you'll be able to better appreciate where it is today in the landscape of artisan alcohol.

You'll also get a taste of the production process to broaden your understanding of the many varieties of rum out there. How different rums are made ties in to how they are best imbibed, whether they're destined to be sipped all on their own or better when blended into daiquiris or mixed into mai tais. Speaking of these classic cocktails, you'll find recipes peppered throughout these pages for making quintessential rum drinks. You'll also find answers to common questions like "What's the difference between rum and rhum?" "What does 'premium' mean anyway?" and "What does the color of rum mean about how it tastes?"

With this richer understanding in your pocket, you'll start to hit the bottle by learning how to decipher liquor labels. Find out what's legalese, what's marketing, what's actually important, and how to tell the difference. Then the best part: really tasting rum. Discover what the experts pay attention to when they're drinking a premium rum so that you can get the most from every sip as you fill out the journal pages with your own tasting notes. From the look and the smell to the taste and the linger, every rum has its own story to tell. You just need to know how to listen.

Finally, you'll visit some of the world's leading rum producers to see what makes them so popular. You can taste along with the notes in this book to get a baseline for rum types and lingo. Your journey through this book will give you a deeper appreciation of the spirit and allow you to fill in your own tasting notes with satisfaction and ease as you go on your own often-tropical adventure over the rainbow of rums.

A BRIEF HISTORY

SWEET BEGINNINGS

Sugarcane has been cultivated since, well, we're not sure when. Some historians say 6000 BCE, others say 8000 BCE. Suffice it to say, it was a *really* long time ago when the Papuans of New Guinea started selectively breeding the sweet stalks of *Saccharum*, which led to the cultivation of *S. officinarum*, the species of sugarcane that would find its way around the globe to become the most popular plant of the world's absolute biggest crop: sugar. But that's later.

The first thing sugarcane did was make its way around the neighboring Pacific islands and onto mainland Asia. The first record of cane being distilled into an "intoxicating drink" is from India in 325 BCE. Distillation was still a long way off, so this drink was a fermented one that was mostly used for medicinal purposes. From there, sugarcane would go to Persia around 600 AD and Europe by way of the Silk Road and the Crusades in the following centuries. The Europeans, in turn, took it with them to the islands off the west coast of Africa at the start of the fifteenth century and then across the Atlantic by the end of it.

Sugar was incredibly valuable at this time of exploration and colonization. It was among the luxury commodities such as spices and coffee that globalization was bringing to European tables. The Portuguese established plantations on Madeira and the Canary

Islands in the mid-fifteenth century. Shortly after that, Christopher Columbus, who had married into a sugar-planter family, brought sugarcane with him for cultivation on his second voyage across the Atlantic in 1493. While cane grew fast in the tropical heat of Brazil and the West Indies, where Spanish, Dutch, French, and English colonizers had started plantations, it also took a lot of hard labor to grow, harvest, and refine the crop into sugar. And that's where things take a turn.

The Dark Side of White Gold

On the one hand, sugar was so valuable in the era of colonialization that it was used as currency in many parts of the newly expanded world. Europeans were crazy for it, and having

it was a luxury and a status symbol. On the other hand, plantation owners were indisputably greedy, and they needed cheap labor to keep their profits up as demand for sugar skyrocketed. Their answer was enslavement and indentured servitude. The infamous "triangle trade" formed as a result: Europe would send its manufactured goods to Africa in exchange for slaves who were sent to the New World to work on plantations that sold their commodities—largely sugar, tobacco, and cotton—to Europe. It was a vicious cycle that endured for centuries.

THE BIRTH OF RUM

As sugar plantations spread across Brazil and the Caribbean islands (Jamaica, Puerto Rico, Cuba, Hispaniola, Martinique, Barbados, and others), so did the problem of molasses. At first, the dark, sticky syrup was seen as just a by-product of refining cane into sugar crystals. Even though there was a lot of sugar in the molasses, there was no good method for converting it into sugar crystals, so it was fed to slaves and livestock, used as fertilizer, and even dumped into the ocean just to get rid of it. Some slaves and peasants fermented it into wine.

Then distilling equipment techniques began arriving in the New World in the early 1600s. At first, French and British plantations started using the sugary scum skimmed off the boiling sugar in the refinery process to make a light sugar spirit until someone thought to try out fermented molasses. The exception to the by-product process was *rhum agricole*, which was made from pressed sugarcane juice on French Martinique starting in the late 1600s. At some point within this murky history, rum as we now know it evolved from experimentation, excess, and necessity. And because rum was associated with industrial waste, essentially, and evolved from a drink favored by slaves and poor people, it wasn't exactly considered upscale. It was popular with middle-class Brits who didn't want to drink gin, which was for poor people closer to home, and couldn't afford brandy. It was also popular with New Englanders.

AN IFFY ETYMOLOGY

The first time "rum," as in the sugarcane spirit, appears in recorded history is on a 1650 deed of sale of a plantation in Barbados. But that wasn't the only word being used for sugarcane spirits from the Caribbean. *Kill devil* was a colorful term used for the spirit, presumably because it was strong enough to kill the devil. It was also called *rumbullion*, an English slang term for "a great tumult or uproar," and that may have been truncated to just *rum*, and certainly rum has been known to cause a tumult or two! But there is some debate if that's the whole story.

Back then, *rum* meant "excellent, fine, or good" and was coupled with *booze* to make *rumbo*, which turned into slang for a strong alcoholic punch, which rum was certainly used for, so perhaps rumbo got shortened to rum, which was then combined

THE SPANISH DID IT FIRST

When people talk about the birth of rum, they typically mean Caribbean rum, which was born of kill devil in the seventeenth century. The Spanish, however, were distilling a sugar spirit called *cachaça* a full century earlier on their plantations in Brazil. The Spanish at this time were fierce isolationists, and they were not widely trading their goods or their expertise with the other colonies, but they did export sugar to the Dutch. It is thought that word of the Spaniards' refinery and distilling techniques may have traveled with the Dutch from Brazil to the Caribbean, but the matter is far from settled.

with the French word *bouillon* meaning "hot drink" and also referred to strong punch to make rumbullion. This had the fun double entendre of an uproar and perhaps that's how it came to be the word that would get shortened to the monosyllabic fun that is rum.

LIFE AND DEATH ON THE HIGH SEAS

However rum was born and named, once it was here, it was here to stay. Its popularity grew tremendously in the seventeenth and eighteenth centuries, as did the murderously unfair economies of the Caribbean. Plantation owners grew richer as they broke the backs of the enslaved people who were forced from their homes in Africa into the bellies of squalid ships and, if they survived the crossing, shackled to a life of hard labor.

The sailors themselves who worked the triangle trade and otherwise spent their lives on the seas drank rum to excess to cope with their own brutal conditions. Sailors in the British Royal Navy were given daily rum rations that, in the 1730s, amounted to half a pint. That's 20 ounces, or just over 13 shots of rum *every day*. A decade later it was mixed with water, limes, and sugar in a concoction known as *grog* to curb the deadly scourge that was scurvy. Vitamin C deficiency among sailors was a gruesome condition that killed more men than battles did. Grog prevented that fate. It also cut down the amount of rum the sailors were drinking, which in turn cut down on the tumultuous uproars that 13 shots a day are bound to cause. The rum ration was cut down and down in the Royal Navy, but it would last until 1972, if you can believe it, when some of the higher-ups decided rum and nuclear subs didn't mix. Sailors were given an extra can of beer to make up for the loss.

This brings us to the most famous rum drinkers of the high seas: pirates! Right? Nay, mateys, nay. It is with heavy heart that I inform you the image of the rum-swigging pirate is a

myth invented by Robert Louis Stevenson in his work of fiction *Treasure Island,* which was published in 1883, a near century after the heyday of privateering. Pirates were partial to brandy and gin, and nobody associated them with rum until Stevenson made up the shanty that appears on the first page of his book: "Fifteen men on the dead man's chest—yo-ho-ho, and a bottle of rum!" Sorry be I.

RUM AND REVOLUTION

North America developed a strong taste for rum very early on, as in pretty much immediately after Europeans started landing there, and certainly long before bourbon was a thing. The first American distillery opened in 1650 in Staten Island, New York, and within a century, there would be at least fifty more throughout New England. Molasses and rum would come up from the Caribbean to meet the rising demands of the increasingly intoxicated northern colonies. Rum was everywhere, and the average adult drank a bottle and a half every week in spiced drinks like mimbo and bombo, mixed with beer to make calibogus and manathan, and caramelized in a foamy drink called flip that was made to bubble using a red-hot poker.

New England colonists also used rum for trade. They exchanged it for furs and food with Native Americans, and they also used it to buy slaves from Africa, which created a smaller triangle trade between New England, Africa, and the Caribbean. This was fine with Mother England when the New Englanders were buying their molasses from British plantations, but as demand grew, American colonists started trading with other countries, that is, the French and the Dutch, who gave them a much better deal. British planters were getting edged out and growing angry about it.

British Parliament passed the Molasses Act of 1733, which taxed the import of molasses from foreign plantations, meaning New England colonists would end up paying the same amount no matter where they bought their molasses. That is, if anyone bothered to pay or enforce the tax. It went largely ignored, so Parliament tried again by passing the Sugar Act of 1764. This tax wasn't as steep but the British were keen on collecting it. The tax dug deep into the profits of New England's booming rum business, and people were . . . unhappy about it. The colonists protested the tax heartily and the government caved. Parliament rolled back the tax in 1766, and the colonists had probably their first tastes of civil disobedience and winning a fight with the crown. Did this help foment the revolution that would start a decade later? Let's just say George Washington insisted on a barrel of rum at his inauguration and leave it there.

BREAKING BONDAGE AND THE SUGAR BEET

Things were changing for the Caribbean in the nineteenth century. The Atlantic slave trade was abolished, which is a priceless advancement for the betterment of humankind. It made sugar production more expensive, but another way of saying that is it made sugar production finally cost something closer to what it's worth, which is money and not human life. Plantations had to start paying their workers and treating them somewhat better, although the pay and the conditions were far from fair.

Another challenge to sugar production was that the soil was becoming exhausted, sapped of its fertility after centuries of overplanting and harvesting. And while sugar production broke down in the islands, it started to pick up in mainland Europe. A method was devised for refining sugar from sugar beets, and by

the end of the century, more than half of the world's sugar was made from beets.

CUBA MAKES THE SCENE

The industrial age revolutionized the way people make just about everything, so too did it change the way rum is distilled. As slavery was being outlawed and beets were taking over the refined sugar industry, distillation technology was finding its way to the shrinking number of distilleries in the Caribbean. New stills were introduced that allowed for continuous distillation, which allowed for more rum to be made more quickly. While the old style of pot stills produced a lighter, some say cleaner rum,

the new continuous stills made rum higher in esters, meaning a darker style that would become known as the Jamaican style.

Scientific experimentation was all the rage in the 1800s, and more and more styles were developed and made replicable. Blending also came into vogue, allowing producers to make the cheap stuff more palatable by adding in a little of the good stuff. The Spanish islands finally lifted their laws against distilling and started to get into the rum game at this time as well. The United States was now a free entity, and trade with it was big business for islands like Cuba. They had nearly 1,400 distilleries in the 1860s, churning out some 5.4 million gallons of rum annually but on the back of slavery, which wouldn't be abolished there until 1886.

Cuban distilleries formulated their own version of rum, which was made lighter to appeal to Americans who were really into martinis at the time and didn't particularly want dark rums. Spanish immigrant Don Facundo Bacardí Massó was the most successful producer of this distinctly Cuban style of rum, taming the yeast and tempering the spirit in oak barrels in the 1860s.

CANTINEROS

Tending bar at the luxurious hotels were the *cantineros*, Cuban bartenders who invented some of the best rum cocktails ever. Think mojitos and daiquiris. An entire cocktail culture sprang up there and roared in the 1920s, while American bars had to speak easy because of Prohibition. Bartending was and continues to be taken very seriously in Cuba. It's a profession you study for and usually stay in for your whole life, as it's well respected and it pays well.

Indeed, Bacardí remains the world's largest rum producer, though it no longer operates out of Cuba, and its signature style is what most Americans think of when they think of rum. Because of political differences, Cuban goods like rum and cigars are still not available for import and sale in the United States. If you want a taste of Cuban rum, you have to go to Cuba yourself, or hope a friend that did brought back a bottle for personal consumption.

A VARIETY OF VARIETIES

Cuba was certainly not the only place to come up with its own style of rum. At the turn of the century, the sugar market was collapsing, thanks largely to the beet technique, so the sugar plantations in the Caribbean turned their focus almost exclusively to rum. Martinique had long been making its own variety, rhum agricole, as did Jamaica. Jamaica, however, was experimenting with different styles to appeal to different markets. It made a dark and heavy rum for the Germans, a light rum to drink on island, and a fruity rum that relied heavily on dunder (see page 35) for export to the UK. Guyana was making its own style that had a short fermentation period. Barbados was making flavored rums. Puerto Rican rum was molasses-heavy with short aging times in oak barrels and double distilled. Distilleries were also now operating in other sugar-growing countries around the world, with some very big operations in places like Australia, the Philippines, and India.

RISE AND FALL AND RISE

World War I was good for the rum business. American whiskey distilleries were commandeered to supply alcohol for munitions, and rum imports filled the empty flasks of Americans at home and on the front. After the war, the temperance movement had its way and Prohibition was instituted in the United States. Rumrunners

brought booze stateside, often watering it down or otherwise adulterating it to turn bigger profits. Americans with the means to do it went to Cuba to party. Havana became a massively popular tourist destination with grand hotels, racetracks, dance halls, golf courses, and of course plenty of rum. When Prohibition was repealed, the tiki craze, with its laid-back ideals about carefree fun in the Hawaiian sun, began in the 1930s. It fueled the craving for tropical drinks in America, and rum was central to fruity concoctions that were often served in coconuts or bowls bigger than your head. The zombie was born, as was the mai tai—both American inventions.

As the twentieth century wore on, and tiki went out with the tide, rum waned in popularity. There were so many kinds of rum that it was sort of hard to tell what rum even was. Not only were different islands making different styles of rum, but different areas of the world had embraced different styles. Americans thought rum was a strong mixer for tropical drinks or to put in Coca-Cola. The UK thought of it as a dark spirit, while its native region of the Caribbean drank a light, clear version of it. The idea of rum had been watered down, and most people took to calling it Bacardí, which tells you something about the kind of rum that was still doing OK.

In the second half of the twentieth century, distillers that managed to stay open were consolidated into larger corporations, including Bacardí itself. Cuba went communist and stopped exporting rum. The distilleries in the US territories of the Virgin Islands and Puerto Rico were kept afloat through tax breaks, and duty-free quotas were instituted by the European Union (EU) for other Caribbean distilleries to keep them going from the 1970s on. The EU dropped the quotas in 1997 but then had to provide a $70 million stimulus package to save the industry four years later. This infusion of money that was meant to help develop distiller-owned brands worked.

Over the next twenty years, as drinkers' interest in craft and small-batch booze rose, as did people's esteem of rum. Rum sales skyrocketed, especially in Asia. Premium brands of long-aged bottles have become increasingly appreciated, sought after, and expensive. There are of course big corporations with large-scale production and distribution—Captain Morgan, Bacardí, and Tanduay, a rum from the Philippines that is in fact the bestselling rum in the world even if you've never heard of it—but there are also many small distilleries that are gaining international acclaim. The heyday of high-end rum has finally dawned.

* * *

MAKING
RUM

SIMPLE PROCESS, ENDLESS POSSIBILITIES

The template for making rum is pretty simple. Ferment a sugar solution using yeast, distill it, give it some time to age, and you've got rum. Pretty straightforward, right? Nope. No way. Within those seemingly simple steps is a vast universe of possibility where every choice, every ingredient, every action, has an effect on the rum. The sugar matters (maybe), the yeast matters, the length of fermentation matters, the way you distill it matters, how long you age it and in what matters, if and how you blend it matters. Every step in the process changes the end result. And since there is no universal standard for what even constitutes calling a spirit rum, let alone how it's produced, rum is made around the world using regional techniques and traditions. Sure, this makes categorization a little complicated—some experts would argue even impossible— but it also means we are blessed with myriad varieties and grades of rum, all with their own distinct flavor and character. From field to bottle, let's take a look at how rum is made. Basically.

OH, SUGAR SUGAR

The moment you cut down a cane, the sugar starts to change. Essentially the sucrose starts to decompose, making it harder to refine the cane into crystallized sugar. That's why most cane is milled within twenty-fours of being cut down. Some places still hand-cut cane using machetes, but in our industrialized world, mostly machines do it for us, working around the clock during the dry season in sugar-producing climes around the world.

At the mills, the canes are cut up further and crushed to release their sweet, sweet cane juice. The juice is then clarified with lime— as in calcium hydroxide, not the fruit used to make mojitos— and heated, which concentrates it into a thin syrup. This syrup goes through a series of evaporators to remove more water and

further concentrate the sugar. Sugar crystals are then seeded into the continually heated supersaturated syrup, which causes mass crystallization as more water evaporates off. This is usually done three times, with the syrup thickening with each pass.

After the first boil, you get cane syrup, also called first syrup or light molasses. After the second boil, you have dark molasses, also called medium molasses, just to keep you on your toes. After the third and final boil you have blackstrap molasses, and that's what's typically used in rum production. Of course, it's not that simple. Some rum is made from cane juice, namely, rhum agricole and cachaça (if you agree that cachaça is indeed rum, which some people don't). Other rum is made using first syrup, while yet others are made with cane honey which is made by concentrating cane juice but never crystallizing it.

Producers who use cane juice and honey to distill their rum are, understandably, more concerned with cane variety and terroir than those who use molasses. The less processed the sugar base, the more the source cane impacts the results, while more-refined (in the literal sense) molasses brings less of its source flavors to bear on the rum, but aspects like pH, ash, and acidity carry more importance.

HAIL TO THE YEAST

Regardless of the sugar source, the yeast used to pull off the somewhat miraculous conversion of sugar into alcohol during fermentation can have a huge impact on the flavor of the rum.

When yeast meets sugar, things really start to cook. The yeast goes to town breaking down the sugar molecules into alcohol while releasing heat and carbon dioxide. Historically, wild yeast in the air was what got this fermentation party started, and while wild yeast is occasionally used today, it's far more common for rum producers to use manufactured yeast. This can be run-of-the-mill commercial dry yeast, or, for the more artisanal lines, proprietary yeast strains are carefully developed and controlled. The yeast is then poured or pumped into open vats of either straight cane juice, or molasses or syrup that's been diluted with water to make it thin enough for the yeast to do its work.

Yeast is both prolific and temperamental. It's a single-cell fungus capable of rampant reproduction and can be very successfully dried and stored. And while you have to heat yeast to get fermentation started, if it gets too hot, around 95°F (35°C), it dies. This is tricky for rum producers in the tropical climes where sugar grows best, as the weather routinely brings temperatures up to and over 95. That means they have to institute some tight temperature controls during fermentation, and the longer the fermentation goes, the longer they have to maintain that control.

Some rum goes through rapid fermentation, taking only a day or two for light rums, whereas other darker rums ferment for many days or even weeks. And the longer the fermentation, the stronger the flavor. This is because, beyond converting sugars into alcohol, the yeast interacts with all sorts of compounds in the sugar base, which then interact with each other. The longer those interactions go on, the more acidity builds up, creating more compounds known as *esters*, which mean more flavor. Any substance produced during fermentation that's not alcohol, including esters, is called a *congener*.

DISTILL MY BEATING HEART

After fermentation, you have what's called a wash. The wash is a diluted solution with a 4 to 9 percent abv (alcohol by volume) and all the flavors and alcohol created during fermentation. To up the level of alcohol by volume, the wash needs to be distilled down, and while rum producers have a lot of options for doing this, here's the basic premise:

Alcohol has a lower boiling point than water, so if you boil the wash in a closed vessel (i.e., not an open vat like the ones used in fermentation), the alcohol will evaporate up faster than the water. The vessel is closed, remember, and the vapor is channeled into a condensing system where it is cooled and turns back into liquid with a higher abv than it had before. The more you vaporize and condense your wash, the more you remove compounds and water, which ups the alcohol by volume and lightens the liquid. To achieve this distillation, you use, you guessed it, a still. There are two main types of stills.

Pot Stills

Call them traditional, call them historical, or call them inefficient, pot stills are the oldest, simplest apparatus for distilling alcohol, and a lot of heavy rum is still made in these. They're kind of like a giant copper kettle and are popularly used in Jamaica and Barbados. Liquid is heating in them so that vapor rises and gets pushed into a neck that angles severely away from the pot. This vapor inevitably has some water in it as well as those flavorful congeners from fermentation, which isn't a bad thing. The taller the still, however, the more compounds get left behind and the lighter the result will be.

From the neck, the vapor goes into condensers where it turns back into a liquid with about 25 percent abv. This is called the *low wine* and, as you can tell by the name, it's not strong enough to be a spirit yet. The low wine is distilled a second time, and this time volatile compounds known as "heads" are "cut" (removed) from the "heart," the rest of the spirit that vaporizes, travels through the neck, and is condensed again. This brings the abv up to about 65 to 75 percent and now the spirit is truly with us.

Sometimes a "retort" is used in lieu of doing a second distillation. A retort is essentially another pot in between the neck and the condenser. After the wash vaporizes and goes through the neck, it goes into a retort of low wine from a previous distillation where it gets a higher abv. The vapor from there goes into yet another retort with high wine, which has a higher abv from a previous distillation, and then that now even higher abv vapor goes on to the condensers. The liquid is then cut up into heads, heart, low wine, and high wine, and goes on to its next step, whether that be waste, retort, or aging.

Column Stills

While the pot still creates booze in batches, which is highly desirable to the artisanal crowd these days, the column still allows for continuous distillation, as in nonstop, 24/7—so long as there's

wash to distill, this baby is chugging. As the pot still is so called because it's shaped like a pot, I'll leave you to guess why a column still is called a column still.

Column stills, also known as continuous stills for obvious reasons, are found in most modern distilleries, but they've been around since the 1830s. They create a cleaner, more neutral spirit than their pot predecessors and are popular in Cuban rum production. There are a number of column-still setups out there, but when it comes to rum, typically a two-column still is used. The first column is the analyzer and the second, the rectifier, and both are segmented into chambers by perforated metal plates. Wash is pumped through a pipe that coils down through the rectifier where it heats up, on its way to the top of the analyzer. The wash makes its way down the analyzer as steam rises from the bottom. The steam catches the alcohol in the falling wash and carries it up and out of the analyzer. It then goes down a pipe to

DUNDER? MUCK!

In Jamaica, some distillers will collect the residue left over in stills after distillation. This "dunder" has lots of dead yeast and is highly acidic, which means it makes great food for yeast and will produce lots of esters in fermentation. Dunder is mixed with molasses, water, and sometimes fruit in "muck pits," so called because the concoction turns into, well, muck. As bacteria breed, the pit takes on a funk that can be smelled from a good ways off, but as long as the muck is monitored so that it doesn't become too acidic and start making ammonia, it is a very tasty ingredient to add to fermentation.

the bottom of the rectifier, where it evaporates up, condenses on a perforated plate, heats up into a vapor again until it hits the next plate, and repeats this process of going from liquid to vapor over and over until it hits the collecting plate, which is set by the distiller to whatever height they deem right. The abv is dictated by the number of plates the rum passes through on its journey. More plates means more abv but less character.

Those are the basics for the most common methods of distillation. There are loads of variations and hybrid stills used to many different ends to make many different kinds of rum. Specialized varieties like rhum agricole, cachaça, and clairin all have their own methods, and you could likely spend a whole lifetime learning about them if you like. Speaking of, let's move on to aging.

REAL MATURE

All rum is aged, but how and for how long, you guessed it, vary widely depending on the region and variety. Some rum is aged in wooden casks or metal tanks, but by and large, it's left to mature in oak barrels, and of that majority, most is aged in old bourbon

barrels. Oak is great because it's liquid-tight but not airtight. Air and alcohol vapor can pass through sturdy oak barrels, but even better, the oak imparts its own flavors to the rum. Bourbon barrels are particularly good because they have a desirable flavor profile and they're constantly becoming available. Bourbon has to be aged in new, charred oak barrels—reuse is not allowed—and rum production is quick to snatch up these flavorful casks.

While there is no global standard for pretty much anything rum related, in most countries, rum has to be aged for at least a year. In Cuba, the minimum is two. Premium bottles can be aged to fifteen years or more. During that time, the barrel breathes in oxygen and releases the more-aggressive alcohol vapors, known as the angels' share. As this maturation goes on, the rum mellows out and the wood lends it color, tannins, and flavor notes like vanilla and spice. The hotter the climate, the faster this happens, so a Caribbean rum will mature at a faster rate than a New England rum, which is a good thing to remember when you're picking out a rum.

BLENDING: EVERYONE DOES ITS

The near final step in production is blending. Almost all rum is blended with other rums of different ages, maybe different styles made in different kinds of stills, aged in different kinds of casks, and hailing from different countries. This blending can create

more consistency from bottle to bottle when done well, but more importantly, it makes the rum more multifaceted, that is, tasty. Pot-still rums are more complex while column-still rums are purer and more delicate. Young rums are fresh while older rums are deep. By blending these different elements together, you can concoct some really special spirits that are increasingly appreciated by serious sippers.

It is up to the blender to balance the flavors and deliver whatever kind of rum it is they are looking to create. Once the blend is right, the rum is filtered and sometimes sweetened, flavored, or even colored all depending on what the producer is looking to put out into the world. After all this, the rum is bottled, labeled, and shipped out for the pleasure and (responsible) consumption of the masses.

✳ ✳ ✳

READING

LABELS

A WIDE WORLD

Organizing rum by type is not an easy job. Rum is made from sugarcane, but that's almost the only thing all rum has in common. It can be made anywhere in the world—there is no international agreement on production methods, aging, or proof. There are common practices, though, and most producers make their rum between 70 and 80 percent abv, age it for some amount of time, and bottle it at between 40 and 45 percent abv. There are also production methods that are favored in different areas.

For instance, the places that were once colonized by the Spanish—Puerto Rico, the Dominican Republic, Colombia, Cuba, and others—are best known for making añejo rums, meaning rums aged in oak barrels. That's a common understanding of the term, however, and not a legal requirement. Añejo rums tend to be on the smooth side, thanks to the barreling. Darker, fuller rums are typical of Caribbean islands where you now find English as the main language: Jamaica, Antigua, the Bahamas, Barbados, and others.

COMMON LABEL LANGUAGE

While certain production methods are indeed favored or even tradition in different areas, any place can make any kind of rum. So, when it comes to understanding what's in the bottle, it's important to understand the descriptions you find most often on labels. Know that these are descriptors largely based on its color, which can be added after aging for effect, so it's also important to find out the age, how a rum was distilled, and from what (sugarcane juice, syrup, or molasses) when you can. More on that later. For now, let's focus on the "types" of rum you see on a lot of labels.

Dark Rums

Usually made from molasses or caramelized sugar, dark rums tend to be aged longer and have stronger, fuller flavors. "Dark" can mean brown, black, and red, and often the color comes from being aged in heavily charred barrels. Dark rum is typically good for sipping all on its own. Most añejo rums are dark rums.

BLACK RUMS

Like everything else in the rum world, "black rum" has no legal definition. Some black rums are aged for years, but typically a black rum is colored with molasses or caramel to give the appearance of strong, long-aged rum, while in fact they are unaged or only briefly aged. You might expect a full flavor profile and robust mouthfeel, but these young rums are in fact light, fiery, and best used in cocktails.

Aged/Gold Rums

Ranging from true gold to amber in color, these rums are most often aged in oak barrels and that's how they get their color. Used bourbon barrels are a popular choice and help smooth out a rum while lending it notes of oak and spice. Some "gold" rums are made so through added coloring like caramel instead of aging, so be sure to check the age on a bottle of gold rum.

White Rums

Also sometimes labeled as "silver" or "light," these rums tend be mild and on the sweet side. They are often filtered after being briefly aged in stainless steel casks, say three to six months when produced in the warmer climes of Puerto Rico, where they are most commonly made. These rums are best suited to light and bright cocktails such as mojitos.

Premium

A premium rum is often made in small, specially blended batches made with an eye toward a great sipping experience. They are more expensive, and as such you're (hopefully) paying for a quality spirit that was carefully produced and blended. These will often come from boutique brands or the high-end of big brands and likely have complex flavor-profiles that developed from a good number of years in the barrel.

Overproof

A rum with an abv over 50 percent is overproof, meaning it's damn strong stuff. These rums can get as high as 75 percent abv and are often unaged and light or clear in color. If they're dark, they're typically called *naval rum* or *navy strength* and likely got their hue from added colorings. Sipping an overproof rum is probably not something you want to do. This stuff is meant to be burned off in a flambé or at the very least dropped in as a floater or mixer in a cocktail.

Spiced

This is exactly what it sounds like: rum that has added spices. Typically, the spices will amplify or mimic the flavors produced from barrel aging, making the rum taste more complex. The base rum used might be gold or white, and the color you get—which can range from light to dark—is often produced by coloring.

Flavored

Like spiced rum, a flavored rum has ingredients added to change the flavor, but in this case, we're talking about coconut, we're talking about mango, we're talking pineapple, lime, banana, and a whole host of other flavors that make the rum well suited to a

---

SPECIALIZED TYPES OF RUM

There are a few "types" of rum that really mean something. If you see the following on a label, you can be pretty sure it means what it says. You might want to give these a sip.

Cachaça is considered a rum in the United States, but other countries consider it a different kind of spirit altogether. It's made from fermented sugarcane juice in Brazil. **Rhum agricole**, as we've discussed, is a rum made from freshly pressed sugarcane juice and typically comes from Caribbean islands with a French colonial history, especially Martinique. Similarly, in Haiti they used nonfermented sugarcane juice to make their clear and revered **clairin. Demerara** isn't a type of rum, but a sugar-producing region on the coast of Guyana. If you see this term on a bottle, you know you'll be getting rum distilled in that country, though it can be in a range of styles.

particular cocktail or mixing in general. The flavors are usually artificial and added after distillation. Technically you *can* sip them, but that's not really what they're made for.

LABEL LEGALESE

While the "type" of rum on a label can mean a spectrum of things, there are some cold, hard facts every liquor label in the United States has to include. The rule of thumb for rum label, or really any label, is that it'll tell you what it's proud of and be specific about this. If a label doesn't name a distiller or just says "natural flavors" without going into any details, those are signs that it doesn't want you to know those details. On the flip side, be wary of long stories that don't actually say anything. A history of brand is meaningless unless the distiller is still using those practices in those places *today*. You need to be savvy about what a bottle is saying between the lines.

The Brand

This is the name the rum is being sold under: Goslings, Bacardí, Captain Morgan, and so on. The brand name is not allowed to lie to you or mislead you about how old the rum is or where it comes from. This is open to interpretation, however, as you can be fairly sure Malibu isn't made in Malibu.

The Type

Pretty much a label only has to tell you it's rum, which in America essentially just means made from fermented juice of a sugarcane. There are lots of "classifications" that we've already discussed that may tell you a bit more or they may mislead you a bit.

Name and Address vs. Origin

You'll always find a name and address on a bottle, but what that name and address is attached to might not tell you much about where the rum came from. You'll get a bottled by such-and-such distilling company and address that can be its business headquarters instead of where the rum was made. Look for "produced in" or "distilled in" or "product of" to get the actual country of origin. If you don't see a line like that, the rum was produced in the United States as a country of origin is required only on imported booze. Be wary of things like "Caribbean style" or "plantation" or other terms that sound rummy but don't actually mean anything.

ABV and Amount

All alcohol has to have its abv and proof on the bottle. In the United States, rum has to be produced at less than 95 percent abv (190 proof) and bottled at no less than 40 percent abv (80 proof). But that's not true around the world, because there is no international legal standard for rum. A bottle will also tell you how much booze is inside, typically in liters.

Additives and Colors

There are regulations about disclosing "coloring materials" and flavoring that are used to produce rum. They really just amount to a label needing to say something like "with natural flavors" or "with caramel."

GOOD SIGNS

There are a few things you can look for on a label that almost always signify quality. And I don't mean the price tag (though a higher price does often mean more effort went into production).

Age Statements

Rum doesn't have to tell you how old it is in the United States, so if you see an age statement, that's a good sign. It means the youngest rum in the bottle has been aged to at least the age listed on the label. Remember, the hotter the climate, the faster aging happens, so a Caribbean rum will mature at a faster rate than a rum from a less tropical locale. There are also a few countries that have their own laws about aging, so you can occasionally glean something about the age from the country of origin. For instance, Cuba has a minimum two-year aging requirement, while Puerto Rico's is one year, and Martinique's rhum vieux agricole is three years. This means not all ages are equal. Also, the number in a name doesn't mean age. That's probably obvious with Barcardí 151 but less so with Ron Matusalem Gran Reserva 15.

Batch Number or Run

If a label tells you what number the bottle is out of a set number of bottles in that batch, or even if it just says the total number of bottles made in its run, that's a good sign of quality. That's not to say you're definitely going to like it, because small-batch bottles often have a strong, distinct taste and can be polarizing in opinion, but you know they're worth trying. Conversely, be wary of labels that just use the words "small batch" or "limited edition" without any clarification or detail. All on their own, they're likely just marketing.

Distillation Specification

A bottle that just says "rum" or "blended rum" probably hasn't gotten any special treatment in production. However, if you see "single pot-still" or "single column-still" on the label, you can be pretty sure the rum was distilled that way at only one distillery. "Single blended rum" can indicate the rum was distilled using both column and pot stills but all from one distillery. These are signifiers of quality and care.

✳ ✳ ✳

TASTING
RUM

BEFORE YOUR POUR

As with fine wine and food, there is an art to tasting rum that will help you savor the experience and get the most out of the tasting. And it starts before you even put your hands on the bottle. When you want to taste a rum—really explore it for its qualities both bold and subtle—there are a few things you can do to ensure the best experience possible.

Glassware

While rum doesn't have its own special variety of tasting glass like cognac and wine do, your choice of glassware is important. The shape of the glass influences how you smell and drink any beverage. For a serious rum tasting, you want a glass that allows the rum to have a good amount of surface area for it to breathe a bit so that it can open up and let off just a little bit of the initial

alcohol hit, and you want the mouth of the glass to be large enough that your nose fits inside it so you can smell while you sip. There are many varieties of glassware that work well for this purpose. The favorite among serious rum tasters is often a simple sherry glass. A small snifter also works well.

Serving

If you really want to taste a rum for all its complexities, it should be served neat: no ice, no mixers, and room temperature. The pour itself should be relatively light. A little goes a long way in a tasting. You can get a good idea of a rum from just a few sips of as little as a quarter ounce. Small pours are particularly important if you plan on tasting several rums in a row or if you're doing a comparison. The more rum you drink, the more impaired your judgment is likely to become, so staying clearheaded for as long as possible will help with the accuracy of your notes.

It's also a good idea to have some water on hand as a palette cleanser. It will help keep your taste buds accurate, especially if you're trying more than one rum. You can add a twist of citrus if you like. Other good palette cleansers include apple slices, bread, unsalted crackers, and citrus sorbet, and they're all made more effective by washing them down with water.

BEFORE YOUR FIRST SIP

After you've poured your chosen rum in your preferred glass, there are some important observations to make before that first sip.

The Color

The first thing you'll notice is the color. Rums come in a variety of colors ranging from crystal-clear white rums to aptly named black rums. A rum's color can come from the barrel it was aged in, and often the longer a rum is aged, the darker it is. This is especially true if you look at the surface of the rum where it meets the glass. For more on rum colors and styles, see page 40.

Its Legs

Now give your rum a nice swirl around the glass. This will help open it up a little more and it will show off its legs, also known as its lacing. This is the rum that takes its time running back down the sides of the glass, showing off the viscosity of the spirit. The rule of thumb is slow legs mean higher alcohol content and fuller body, but of course there are exceptions to every rule.

The Nose

The next thing you'll want to do is get nosy with your rum. Nosing a rum is alcohol-speak for smelling it with purpose. Take a few short sniffs while passing the rum under your nose, going past both nostrils a couple times. This is the "top sniff" and it will allow you to take in the most obvious aromatic notes the rum has to offer. You can then swirl the rum and stick your nose into the glass to get a whiff of the more subtle and complex aromas.

Because there are so many types of rum from all over the world, there is a wide spectrum of aromatic notes that you may experience. Since rum is made from sugar, the range of sweet smells is particularly robust. Here are just some of the most common, some of which may themselves sometimes have a hint of being burnt, baked, or roasted:

SWEET
caramel/toffee
brown sugar
molasses
honey
nougat
maple syrup
butterscotch
cola
custard
marzipan
marshmallow
chocolate
vanilla
fudge
gingerbread
fruitcake
butter

FRUITY
apple
apricot
cherry
dark fruit
currant
lime
lemon
orange
peach
banana
coconut
mango
pineapple

SPICY
nutmeg
cinnamon
clove
anise
cardamom
vanilla
black pepper
ginger

WOODY AND NUTTY
oak
cedar
sandalwood
almond
walnut
hazelnut
coffee
pecan

EARTHY AND VEGETAL
flowers
grass
herbs
mint
peat
tobacco
roots
musty
tannins
leather
salt
medicine

metal
tar
rubber
char
smoke

ALCOHOLIC
whiskey
sherry
brandy
cognac
wine

TASTING

You've already had a thorough sensory experience with your rum and you haven't even taken your first sip. Your brain and your palate are now primed to drink in the complexities of this diverse spirit. Unlike with some other spirits that smell one way and taste another, rum's taste often matches its aroma. So, if you get leather, caramel, and oak on your sniff test, you're likely to get all or at least some of those flavors on your palate. They may be stronger or subtler, and you may get new flavors as well. You'll also experience how the rum feels in your mouth, including its viscosity, its heat, and how smooth (or harsh) it is.

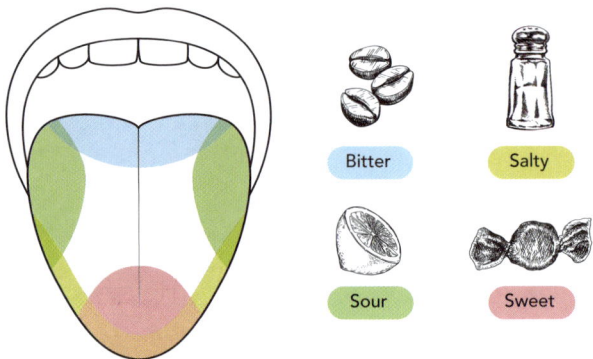

The First Sip

The first sip is, funny enough, the least important. The initial punch of the alcohol often dominates the experience, and you can't yet taste the full depth of flavor. Different areas of the tongue detect different flavors, so be sure your sip travels across your entire tongue. The tip and to some extent the center of your tongue are best at tasting sweetness, so those regions are the most important for this sugar-based spirit.

The Second Sip

Your second sip will give you a deeper sense of the rum's flavor. This time, take a slower sip that's a bit larger than the first. Allow the aroma to waft into your nose as you do so. Really give the rum a chance to travel through your mouth, perhaps giving it a slow swirl so that it hits all the different regions of taste buds. Now you'll likely taste the flavors you detected in the rum's aroma, the molasses, the mahogany, the leather, the figs, whatever they are. But now you'll also get the mouthfeel—literally just how it feels in your mouth—which has several categories of its own.

BODY (VISCOSITY)	BITE
buttery	smooth
creamy	warm
heavy	crisp
syrupy	dry
cloying	harsh
thick	burn
oily	heat
silky	sharp
velvety	astringent
full	
medium	
light	

Notice how the flavors and mouthfeel marry up. Does this create a balanced experience, or are elements of the taste fighting each other? Does one aspect dominate all the others?

The Fade

The last thing to notice is how a rum fades, or doesn't, from your palate after you've swallowed it. Do the flavors linger? Do they change? How long does it take the taste to fade away? Do you feel the need to sip some water?

Subsequent Sips

As you continue to sip the rum, you may notice that some of the flavors and the bite change as you drink it. Some aspects may become more pronounced as your palate becomes more accustomed to them. You might get more of a burn, or the caramel might really come through. Other

aspects might ebb. A heated bite might mellow into warmth or a light body might seem to thicken as you drink.

You might also decide to add a splash of cool water or an ice cube after your first couple of sips. This will of course dilute the rum, making it less intense, which may allow you to detect some of its subtler characteristics.

A TASTING FLIGHT

If you are doing a tasting flight or sampling more than one or two rums, as previously mentioned, it's important to have some water and palate cleansers on hand. You might also want to consider the order in which you're tasting the rums. It is recommended to start with the simplest rum, which is often the youngest and lightest, and going through to the most complex, which is typically the oldest. By going in this order, from easiest to strongest, you preserve your palate and allow it to taste the potentially subtle differences between each rum, instead of going big in the beginning, which has a greater effect on your palate.

FAMOUS RUMS

NAMES YOU KNOW

There is an incredible number of rums out there for you to try, from unaged to decades old, from crystal clear to black as the richest cup of coffee. While more and more small-batch, high-end rums that are certainly worth your attention are finding their way to shelves, there are some nationally distributed brands that everyone knows and form the bedrock of the rum landscape. These are the big boys that you see in every store and every bar, and because they're so consistent and well known, starting your tasting journey with these can give you a solid foundation of the rum basics across a few different styles. This doesn't mean they're the best rums, just the most ubiquitous. Though being widespread doesn't necessarily make them bad.

There is, of course, near endless debate over which rums are the best, which expensive bottles are worth their price tags, and

whether or not some of the most popular brands should just be used as mixers. You'll have to form your own opinions about these issues. What's presented here is the lowdown on a shortlist of some of the most popular— meaning bestselling—rum brands in the United States. While your personal go-to favorite might not be listed here, that just means you can start your tasting notes with that one on the journal pages that follow.

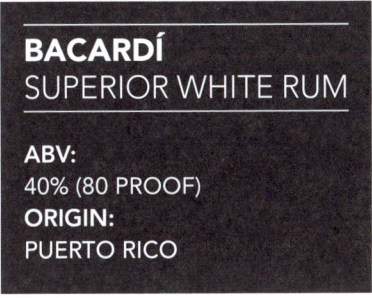

BACARDÍ
SUPERIOR WHITE RUM

ABV:
40% (80 PROOF)
ORIGIN:
PUERTO RICO

Bacardí is—or at least was—practically synonymous with rum for a good part of the late twentieth century. With roots that go back to the 1830s in Santiago, Cuba, Bacardí was started by a Spanish immigrant, Don Facundo Bacardí Massó, who made pioneering innovations in yeast cultivation and charcoal filtration. It became a family business as Facundo's rum found wide commercial success.

Operations in Santiago expanded, and the company started to find success exporting. It began distilling rum in other countries, including Mexico and Puerto Rico. After the Cuban Revolution came to a close in the late 1950s, the new government started seizing some of the company's assets and the business moved to Bermuda, where it is still headquartered. Today, Bacardí Superior is the flagship rum of this expansive spirit company. It is produced based on a recipe created by Facundo and perfected at the

original distillery in Santiago using bourbon barrels for aging and charcoal filtration.

Best Served

Mixed in classic rum cocktails such as mojitos and daiquiris

Tasting Notes

This crystal-clear rum has little legs in the glass. The nose is a bit alcoholic but with notes of almond, lime, and caramel. The initial taste is harsh but sweet. The almond comes through as does some vanilla from the barrel aging, but it dries up pretty quickly. There's also some bananas and a somewhat industrial smoke taste that makes it apparent why this is considered a mixing rum and not a sipping rum. It's certainly light bodied, as white rums are meant to be. It's dry on the finish and somewhat acidic.

CAPTAIN MORGAN
ORIGINAL SPICED
RUM

ABV:
35% (70 PROOF)
ORIGIN:
US VIRGIN ISLANDS

Though produced in the US Virgin Islands today, Captain Morgan was conceived by the Canadian Seagram Company in 1944. Originally distilled in Jamaica using a recipe a Seagram CEO bought from a Kingston pharmacy that was adding medicinal herbs and spices to raw rum, it's been a spiced rum since the start. As the brand grew, operations moved to Puerto Rico where they stayed for decades. The British multinational beverage company Diageo bought the brand in

2001. It announced plans to open a distillery in St. Croix and, a decade later, did so. Captain Morgan himself was indeed a very real historical figure. While the actual privateer Morgan was definitely a bad, bad man, the mascot version of him today is a whimsical rascal who would never invade your house and string you up by your nethers.

Best Served

Mixed with cola or blended in a piña colada

Tasting Notes

This light amber-gold-colored rum has a relatively low abv. At 35 percent it's technically below the threshold for rum in the United States, though since it's "spiced rum" it gets a pass. It has a sweet vanilla, butterscotch aroma with the vanilla notes deepening as it breathes. Presumably that's from the charred white-oak barrel aging and not the "other natural flavors" mentioned on the label. While the first taste you get is pretty astringent, it smooths out quickly and sprints toward a maple finish. It's got a bit of a burn going down, but it doesn't linger long or offend. A considered second sip presents more spiciness and citrus, but that still doesn't really make it great for drinking neat. Best add it to some cola as pretty much all the advertising advises.

MOUNT GAY ECLIPSE

ABV:
40% (80 PROOF)
ORIGIN:
BARBADOS

This is the flagship rum from one of the oldest distilleries in the world. Mount Gay can trace its history to at least 1703, and Eclipse dates back to 1910, a year that had a total eclipse. At that time, the distillery had only one copper still pot, but today, like pretty much all rum, Eclipse is a blend. Some of the rum that goes into it is distilled in copper still pots, while the rest is made in a two-column copper still. Most of the sugar used by Mount Gay is locally grown, and the resulting molasses is mixed with water from a coral limestone well on its property in northern Barbados.

Mount Gay also has a proprietary yeast strain and ferments Eclipse in oak vats. It is then aged in oak bourbon barrels. So, while the operation is quite big, it's a careful process that's not exactly what you'd think of as industrial.

Best Served
With cola or in punch but certainly not bad on its own

Tasting Notes
This gold rum has a hint of amber in its appearance with sturdy, slow legs. Brown sugar and orange are what you mostly get from the nose, although there's some spiciness to it as well. This all carries over to

the flavor, though leaving a lot of the sweetness behind. You also get some dried fruits and tobacco, perhaps a hint of leather in a fairly complex blend. Dry with a bit of a spicy burn during a medium finish, it's a solid rum that does all the things rum should do.

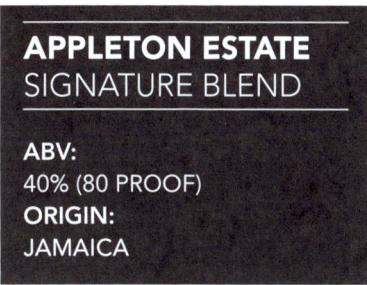

APPLETON ESTATE
SIGNATURE BLEND

ABV:
40% (80 PROOF)
ORIGIN:
JAMAICA

The Appleton Estate (an actual estate, not just a marketing term) is located in the fertile Nassau valley of Jamaica. The estate was purchased by J. Wray & Nephew Ltd. in 1916 and then the Campari Group in 2012, but rum has been made there since 1749. The estate makes rum from "cane to cup" as it grows, mills, and distills there, under the direction of Joy Spence, the world's first female master blender. Its Signature Blend is an añejo (aged) rum. Though it doesn't provide an age statement, it's made from fifteen rums that go up to twelve years old but average four. Made in small batches in pot stills and aged in oak barrels, it's an iconic Jamaican rum with a deep history.

Best Served

Mixed in citrusy drinks like daiquiris and mai tais or sipped straight

Tasting Notes

The color of amber honey, this rum has some long legs. The nose is boldly molasses with fermented fruits including banana and orange and a little black pepper. The orange and spice come through in the taste with some ginger and woodiness joining them. Brown sugar also carries through. It's a hot and dry rum on further sips, but not rough, and it has an impressively spicy finish that still brings the orange and oak with it.

EL DORADO
12 YEAR OLD RUM

ABV:
40% (80 PROOF)
ORIGIN:
GUYANA

OK, this might not be in *every* bar and *every* store, but it's not exactly rare either. El Dorado is the flagship brand of Demerara Distillers on the coast of Guyana where sugar and rum have been in production since the 1600s.

While hundreds of years ago there were hundreds of distilleries, today, this is the only one still operating in the country, having consolidated all the smaller operations into its fold. As such, it has some unique distilling operations, such as the world's only wooden continuous still and the last two wooden pot stills in the world. Demerara matures its rum in oak barrels and is known for long aging periods, as is evident with its 25-year-old rum. Its 12-year-old blend is a good example of what its old-school operations yield without going too high on the price point.

Best Served

Straight or on the rocks, though certainly wouldn't ruin a cocktail

Tasting Notes

Though not as dark as it looks in the slightly tinted old-timey bottle, this rum is deep gold edging into dark with legs that are slow to form. Dark fruits as well as some citrus rise in the aroma along with dark-brown sugar and some baking spices as it airs out a bit. Medium- to full-bodied but mellow, the flavor balances sweetness, dark fruit, vanilla, oak, those baking spices again (especially cinnamon), and tobacco. It

has a long, somewhat dry finish in which the dark fruits linger. It's certainly complex yet smooth and easy to sip, which allows you to spend time pondering all the flavors.

✳ ✳ ✳

RUM
COCKTAILS

DAIQUIRI

Perhaps *the* quintessential rum cocktail from Cuba, the daiquiri has a long, partially uncertain history. Suffice it to say that a manager of a Spanish-American mining operation in Daiquirí, Cuba, was very fond of a drink that was essentially a rum sour that used brown sugar and fresh-squeezed lime. Somehow or other, this drink got the name of the town. From there it traveled to Havana, where it was perfected over decades by talented cantineros, namely Constante Ribalaigua Vert and Miguel Boadas at the Floridita, which has come to be known as *cuna de la daiquiri* ("cradle of the daiquiri").

2 ounces white rum
½ ounce lime juice, fresh
  squeezed

1 teaspoon white sugar

Add all ingredients to a shaker full of ice and shake vigorously. Like a *lot*. Strain into a cocktail glass. That's it!

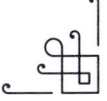

MAI TAI

The story of the mai tai goes back to 1944 when Trader Vic Bergeron, an influential entrepreneur of the tiki craze, mixed up a drink for a couple visiting his then-famous bar. He used seventeen-year-old Wray & Nephew rum (today Appleton Estates) in a tropical-inspired concoction that the woman liked so much, she yelled, "Maita'i!" which is Tahitian for "good."

While this is a feel-good story for a drink practically synonymous with relaxation, the rest of the history is anything but. A fight broke out between Vic and his rival, Donn "the Beachcomber" Beach, who claimed the mai tai was a rip-off of his Q.B. Cooler. This ended in a lawsuit that forced Vic to reveal his once-secret recipe.

2 ounces aged Jamaican rum
½ ounce curaçao orange
  liqueur
½ ounce orgeat syrup
½ ounce simple syrup

1 ounce lime juice, fresh
  squeezed
crushed ice
mint for garnish

Fill a shaker with ice and pour in all the liquid ingredients. Shake until well chilled and pour into a double old-fashioned or tall glass. Fill with crushed ice. Smack the mint and add as garnish.

VARIATION

There are those who would have you use 1 ounce aged Jamaican rum and 1 ounce rhum agricole to better approximate the seventeen-year-old Wray & Nephew.

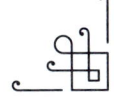

PIÑA COLADA

In the early days of the piña colada, there was no coconut in it. Just fresh pineapple juice and aguardiente (Spanish schnapps), ice chilled. The tiki craze caused coconut cream to be manufactured and sold in tins. This was added to the cocktail in the 1950s. After that, it was all downhill as premade mixes and slushy machines became the main way this drink is made. While those noxious chemical concoctions are good only because you typically get them pool- or beachside, piña coladas can be made enjoyable regardless of the scenery.

1 ounce white rum
1 ounce gold rum
2 ounces pineapple juice

1 ounce coconut cream
crushed ice
1 pineapple wedge for garnish

In a blender, blitz together all the ingredients with roughly a cup of crushed ice or small ice cubes (the big ones take too long to blend). Pour into a tall glass such as a Collins or hurricane glass. Garnish with your wedge.

RUM PUNCH

Back in the 1700s, rum was a middle-class drink for the English. The rich could afford cognac and the poor swilled rum, so the middle class took to rum, which had the status of West Indian trading but was still somewhat affordable. Rum punch became popular in gentleman clubs and public houses, some of which dealt so heavily in punch that they became known as "punch houses." Back in that heyday, there was a big punch bowl that patrons could go to to fill their cups. As time went on, the bowl disappeared from clubs and went behind the bar where the libation evolved into the cocktail we can't agree on today. The variations are endless and up to your tastes. Here's this author's favorite.

1 shot dark rum
3 ounces pineapple juice
2 ounces guava juice

1 lime wedge
nutmeg

Fill a mixing glass with ice and pour in the liquid ingredients. Squeeze the juice from the lime wedge. Stir until well chilled. Pour into a hurricane or double old-fashioned glass filled with ice. Sprinkle a light dusting of nutmeg over top.

VARIATIONS

As I said, they're endless. Grenadine or orange juice in lieu of guava juice is pretty popular. If you want to scale this up and serve it in a punch bowl, you can mix it right in the bowl with ice to keep things chill.

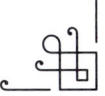

RUM & SODA

Some of the best-known rum drinks are as simple as rum + soda poured over ice with a lime wedge. Here are the best ratios for two of these popular rum drinks.

CUBA LIBRÉ (AKA RUM & COKE)

When bottles of Coca-Cola started arriving in Cuba in the early 1900s, adding rum to it was a natural next step. The drink was named *Cuba libré* ("free Cuba"), the slogan for Cuban independence. Bacardí advertising executive Fausto Rodriguez signed an affidavit swearing he was there the very first time rum met Coke. It was in August 1900 at a bar in Havana. The rum in question? Of course it was Bacardí.

1 shot white rum
3 ounces Coke

1 lime wedge

Fill a tall glass with ice. Pour in the rum, then add the Coke and squeeze that wedge and drop it in. Optional: Stir using a barspoon. Note: The ratio of rum to coke is 1:2. You can scale that however you like.

DARK 'N' STORMY

This gingery rum drink is indeed a classic, dating back to the colonial days in Bermuda. The British Gosling family established a rum distillery on the island in 1857, and soon after people began mixing their Black Seal rum with the ginger beer being made at the nearby Royal Naval Officer's Club. It was a match made in paradise.

2 ounces dark rum
4 ounces ginger beer

1 lime wedge

Fill a tall glass with ice. Pour in the rum, then add the ginger beer and squeeze in some fresh lime juice. Stir using a barspoon and garnish with the lime wedge, by either dropping in the squeezed wedge or perching a fresh one on the rim of the glass.

TASTING NOTES

RUM
The Details

Brand Name

Producer | **Place of Origin** | **Price per Glass/Bottle**

Distiller | **Cask Type** | **Date**

Age | **ABV/Proof** | **Tasting Location**

NOSE
Choose All That Apply

SWEET
- ○ caramel/toffee
- ○ brown sugar
- ○ molasses
- ○ honey
- ○ maple syrup
- ○ chocolate
- ○ cola
- ○ butterscotch
- ○ fruitcake
- ○ gingerbread

FRUITY
- ○ apple
- ○ apricot
- ○ cherry
- ○ currant
- ○ dark fruit
- ○ lemon
- ○ lime
- ○ orange
- ○ peach
- ○ banana

- ○ coconut
- ○ mango
- ○ pineapple

WOODY AND NUTTY
- ○ oak
- ○ cedar
- ○ sandalwood
- ○ coffee
- ○ almond
- ○ walnut

- ○ hazelnut
- ○ pecan

EARTHY AND VEGETAL
- ○ flowers
- ○ grass
- ○ herbs
- ○ tobacco
- ○ tannins
- ○ leather
- ○ salt

- ○ medicine
- ○ metal
- ○ smoke

ALCOHOLIC
- ○ whiskey
- ○ sherry
- ○ brandy
- ○ cognac
- ○ wine

OTHER

STYLE
Choose One

- ○ Neat
- ○ With Water
- ○ On the Rocks
- ○ Cocktail

COLOR
Circle One

TASTING WHEEL

Rate your tasting experience, 1 (lowest) to 5 (highest)

Legs, Sweetness, Molasses, Fruity, Woody, Nutty, Earthy, Vegetal, Spicy, Leathery, Smoky, Bite, Heat, Body, Balance, Linger

Describe the First Sip

....
....
....

Describe the Third Sip

....
....
....

Describe the Fade

....
....
....

What Is Most Striking About This Rum?

....
....
....

Additional Notes

....
....
....
....

GUIDED TASTING
Write It Out

QUALITY RATING

COST RATING

OVERALL RATING

RATE IT
Fill It In

RUM

The Details

Brand Name

Producer

Place of Origin

Price per Glass/Bottle

Distiller

Cask Type

Date

Age

ABV/Proof

Tasting Location

NOSE

Choose All That Apply

SWEET
- ○ caramel/toffee
- ○ brown sugar
- ○ molasses
- ○ honey
- ○ maple syrup
- ○ chocolate
- ○ cola
- ○ butterscotch
- ○ fruitcake
- ○ gingerbread

FRUITY
- ○ apple
- ○ apricot
- ○ cherry
- ○ currant
- ○ dark fruit
- ○ lemon
- ○ lime
- ○ orange
- ○ peach
- ○ banana

- ○ coconut
- ○ mango
- ○ pineapple

WOODY AND NUTTY
- ○ oak
- ○ cedar
- ○ sandalwood
- ○ coffee
- ○ almond
- ○ walnut

- ○ hazelnut
- ○ pecan

EARTHY AND VEGETAL
- ○ flowers
- ○ grass
- ○ herbs
- ○ tobacco
- ○ tannins
- ○ leather
- ○ salt

- ○ medicine
- ○ metal
- ○ smoke

ALCOHOLIC
- ○ whiskey
- ○ sherry
- ○ brandy
- ○ cognac
- ○ wine

OTHER

STYLE

Choose One

- ○ Neat
- ○ With Water
- ○ On the Rocks
- ○ Cocktail

COLOR

Circle One

TASTING WHEEL

Rate your tasting experience, 1 (lowest) to 5 (highest)

Legs, Sweetness, Molasses, Fruity, Woody, Nutty, Earthy, Vegetal, Spicy, Leathery, Smoky, Bite, Heat, Body, Balance, Linger

Describe the First Sip

....

....

....

Describe the Third Sip

....

....

....

Describe the Fade

....

....

....

What Is Most Striking About This Rum?

....

....

....

Additional Notes

....

....

....

....

GUIDED TASTING

Write It Out

QUALITY RATING

COST RATING

OVERALL RATING

RATE IT

Fill It In

RUM
The Details

Brand Name

Producer

Place of Origin

Price per Glass/Bottle

Distiller

Cask Type

Date

Age

ABV/Proof

Tasting Location

NOSE
Choose All That Apply

SWEET
- ○ caramel/toffee
- ○ brown sugar
- ○ molasses
- ○ honey
- ○ maple syrup
- ○ chocolate
- ○ cola
- ○ butterscotch
- ○ fruitcake
- ○ gingerbread

FRUITY
- ○ apple
- ○ apricot
- ○ cherry
- ○ currant
- ○ dark fruit
- ○ lemon
- ○ lime
- ○ orange
- ○ peach
- ○ banana

- ○ coconut
- ○ mango
- ○ pineapple

WOODY AND NUTTY
- ○ oak
- ○ cedar
- ○ sandalwood
- ○ coffee
- ○ almond
- ○ walnut

- ○ hazelnut
- ○ pecan

EARTHY AND VEGETAL
- ○ flowers
- ○ grass
- ○ herbs
- ○ tobacco
- ○ tannins
- ○ leather
- ○ salt

- ○ medicine
- ○ metal
- ○ smoke

ALCOHOLIC
- ○ whiskey
- ○ sherry
- ○ brandy
- ○ cognac
- ○ wine

OTHER
....

STYLE
Choose One

- ○ Neat
- ○ With Water
- ○ On the Rocks
- ○ Cocktail

TASTING WHEEL

Rate your tasting experience, 1 (lowest) to 5 (highest)

Legs, Sweetness, Molasses, Fruity, Woody, Nutty, Earthy, Vegetal, Spicy, Leathery, Smoky, Bite, Heat, Body, Balance, Linger

COLOR
Circle One

TASTING NOTES

Describe the First Sip

....

....

....

Describe the Third Sip

....

....

....

Describe the Fade

....

....

....

What Is Most Striking About This Rum?

....

....

....

Additional Notes

....

....

....

....

GUIDED TASTING

Write It Out

QUALITY RATING

COST RATING

OVERALL RATING

RATE IT

Fill It In

ENJOYING RUM

RUM
The Details

Brand Name

Producer

Place of Origin

Price per Glass/Bottle

Distiller

Cask Type

Date

Age

ABV/Proof

Tasting Location

NOSE
Choose All That Apply

SWEET
- ○ caramel/toffee
- ○ brown sugar
- ○ molasses
- ○ honey
- ○ maple syrup
- ○ chocolate
- ○ cola
- ○ butterscotch
- ○ fruitcake
- ○ gingerbread

FRUITY
- ○ apple
- ○ apricot
- ○ cherry
- ○ currant
- ○ dark fruit
- ○ lemon
- ○ lime
- ○ orange
- ○ peach
- ○ banana

- ○ coconut
- ○ mango
- ○ pineapple

WOODY AND NUTTY
- ○ oak
- ○ cedar
- ○ sandalwood
- ○ coffee
- ○ almond
- ○ walnut

- ○ hazelnut
- ○ pecan

EARTHY AND VEGETAL
- ○ flowers
- ○ grass
- ○ herbs
- ○ tobacco
- ○ tannins
- ○ leather
- ○ salt

- ○ medicine
- ○ metal
- ○ smoke

ALCOHOLIC
- ○ whiskey
- ○ sherry
- ○ brandy
- ○ cognac
- ○ wine

OTHER

STYLE
Choose One

- ○ Neat
- ○ With Water
- ○ On the Rocks
- ○ Cocktail

COLOR
Circle One

TASTING WHEEL

Rate your tasting experience, 1 (lowest) to 5 (highest)

Legs · Sweetness · Molasses · Fruity · Woody · Nutty · Earthy · Vegetal · Spicy · Leathery · Smoky · Bite · Heat · Body · Balance · Linger

TASTING NOTES

Describe the First Sip

....

....

....

Describe the Third Sip

....

....

....

Describe the Fade

....

....

....

What Is Most Striking About This Rum?

....

....

....

Additional Notes

....

....

....

....

QUALITY RATING	COST RATING	OVERALL RATING

RUM
The Details

Brand Name

Producer **Place of Origin** **Price per Glass/Bottle**

Distiller **Cask Type** **Date**

Age **ABV/Proof** **Tasting Location**

NOSE
Choose All That Apply

SWEET
- caramel/toffee
- brown sugar
- molasses
- honey
- maple syrup
- chocolate
- cola
- butterscotch
- fruitcake
- gingerbread

FRUITY
- apple
- apricot
- cherry
- currant
- dark fruit
- lemon
- lime
- orange
- peach
- banana

- coconut
- mango
- pineapple

WOODY AND NUTTY
- oak
- cedar
- sandalwood
- coffee
- almond
- walnut

- hazelnut
- pecan

EARTHY AND VEGETAL
- flowers
- grass
- herbs
- tobacco
- tannins
- leather
- salt

- medicine
- metal
- smoke

ALCOHOLIC
- whiskey
- sherry
- brandy
- cognac
- wine

OTHER

STYLE
Choose One

- Neat
- With Water
- On the Rocks
- Cocktail

TASTING WHEEL

Rate your tasting experience, 1 (lowest) to 5 (highest)

Tasting wheel with spokes labeled: Legs, Sweetness, Molasses, Fruity, Woody, Nutty, Earthy, Vegetal, Spicy, Leathery, Smoky, Bite, Heat, Body, Balance, Linger. Each spoke numbered 1 (center) to 5 (outer).

COLOR
Circle One

TASTING NOTES

Describe the First Sip

....

....

....

Describe the Third Sip

....

....

....

Describe the Fade

....

....

....

What Is Most Striking About This Rum?

....

....

....

Additional Notes

....

....

....

....

GUIDED TASTING

Write It Out

QUALITY RATING

COST RATING

OVERALL RATING

RATE IT

Fill It In

RUM
The Details

Brand Name

Producer   Place of Origin   Price per Glass/Bottle

Distiller   Cask Type   Date

Age   ABV/Proof   Tasting Location

NOSE
Choose All That Apply

SWEET
- caramel/toffee
- brown sugar
- molasses
- honey
- maple syrup
- chocolate
- cola
- butterscotch
- fruitcake
- gingerbread

FRUITY
- apple
- apricot
- cherry
- currant
- dark fruit
- lemon
- lime
- orange
- peach
- banana

- coconut
- mango
- pineapple

WOODY AND NUTTY
- oak
- cedar
- sandalwood
- coffee
- almond
- walnut

- hazelnut
- pecan

EARTHY AND VEGETAL
- flowers
- grass
- herbs
- tobacco
- tannins
- leather
- salt

- medicine
- metal
- smoke

ALCOHOLIC
- whiskey
- sherry
- brandy
- cognac
- wine

OTHER

STYLE
Choose One

- Neat
- With Water
- On the Rocks
- Cocktail

TASTING WHEEL
Rate your tasting experience, 1 (lowest) to 5 (highest)

COLOR
Circle One

TASTING NOTES

Describe the First Sip

....

....

....

Describe the Third Sip

....

....

....

Describe the Fade

....

....

....

What Is Most Striking About This Rum?

....

....

....

Additional Notes

....

....

....

....

GUIDED TASTING

Write It Out

QUALITY RATING

COST RATING

OVERALL RATING

RATE IT

Fill It In

RUM
The Details

Brand Name

Producer **Place of Origin** **Price per Glass/Bottle**

Distiller **Cask Type** **Date**

Age **ABV/Proof** **Tasting Location**

NOSE
Choose All That Apply

SWEET
- ○ caramel/toffee
- ○ brown sugar
- ○ molasses
- ○ honey
- ○ maple syrup
- ○ chocolate
- ○ cola
- ○ butterscotch
- ○ fruitcake
- ○ gingerbread

FRUITY
- ○ apple
- ○ apricot
- ○ cherry
- ○ currant
- ○ dark fruit
- ○ lemon
- ○ lime
- ○ orange
- ○ peach
- ○ banana

- ○ coconut
- ○ mango
- ○ pineapple

WOODY AND NUTTY
- ○ oak
- ○ cedar
- ○ sandalwood
- ○ coffee
- ○ almond
- ○ walnut

- ○ hazelnut
- ○ pecan

EARTHY AND VEGETAL
- ○ flowers
- ○ grass
- ○ herbs
- ○ tobacco
- ○ tannins
- ○ leather
- ○ salt

- ○ medicine
- ○ metal
- ○ smoke

ALCOHOLIC
- ○ whiskey
- ○ sherry
- ○ brandy
- ○ cognac
- ○ wine

OTHER

STYLE
Choose One

- ○ Neat
- ○ With Water
- ○ On the Rocks
- ○ Cocktail

COLOR
Circle One

TASTING WHEEL

Rate your tasting experience, 1 (lowest) to 5 (highest)

Tasting wheel with axes: Legs, Sweetness, Molasses, Fruity, Woody, Nutty, Earthy, Vegetal, Spicy, Leathery, Smoky, Bite, Heat, Body, Balance, Linger — each scaled 1 to 5.

TASTING NOTES

Describe the First Sip

....

....

....

Describe the Third Sip

....

....

....

Describe the Fade

....

....

....

What Is Most Striking About This Rum?

....

....

....

Additional Notes

....

....

....

....

QUALITY RATING	COST RATING	OVERALL RATING

RUM
The Details

Brand Name		
Producer	**Place of Origin**	**Price per Glass/Bottle**
Distiller	**Cask Type**	**Date**
Age	**ABV/Proof**	**Tasting Location**

NOSE
Choose All That Apply

SWEET
- ○ caramel/toffee
- ○ brown sugar
- ○ molasses
- ○ honey
- ○ maple syrup
- ○ chocolate
- ○ cola
- ○ butterscotch
- ○ fruitcake
- ○ gingerbread

FRUITY
- ○ apple
- ○ apricot
- ○ cherry
- ○ currant
- ○ dark fruit
- ○ lemon
- ○ lime
- ○ orange
- ○ peach
- ○ banana

- ○ coconut
- ○ mango
- ○ pineapple

WOODY AND NUTTY
- ○ oak
- ○ cedar
- ○ sandalwood
- ○ coffee
- ○ almond
- ○ walnut

- ○ hazelnut
- ○ pecan

EARTHY AND VEGETAL
- ○ flowers
- ○ grass
- ○ herbs
- ○ tobacco
- ○ tannins
- ○ leather
- ○ salt

- ○ medicine
- ○ metal
- ○ smoke

ALCOHOLIC
- ○ whiskey
- ○ sherry
- ○ brandy
- ○ cognac
- ○ wine

OTHER

STYLE
Choose One

- ○ Neat
- ○ With Water
- ○ On the Rocks
- ○ Cocktail

COLOR
Circle One

TASTING WHEEL

Rate your tasting experience, 1 (lowest) to 5 (highest)

TASTING NOTES

Describe the First Sip

....

....

....

Describe the Third Sip

....

....

....

Describe the Fade

....

....

....

What Is Most Striking About This Rum?

....

....

....

Additional Notes

....

....

....

....

QUALITY RATING   **COST RATING**   **OVERALL RATING**

RATE IT

Fill It In

RUM
The Details

Brand Name

Producer	**Place of Origin**	**Price per Glass/Bottle**
Distiller	**Cask Type**	**Date**
Age	**ABV/Proof**	**Tasting Location**

NOSE
Choose All That Apply

SWEET
- ○ caramel/toffee
- ○ brown sugar
- ○ molasses
- ○ honey
- ○ maple syrup
- ○ chocolate
- ○ cola
- ○ butterscotch
- ○ fruitcake
- ○ gingerbread

FRUITY
- ○ apple
- ○ apricot
- ○ cherry
- ○ currant
- ○ dark fruit
- ○ lemon
- ○ lime
- ○ orange
- ○ peach
- ○ banana

- ○ coconut
- ○ mango
- ○ pineapple

WOODY AND NUTTY
- ○ oak
- ○ cedar
- ○ sandalwood
- ○ coffee
- ○ almond
- ○ walnut

- ○ hazelnut
- ○ pecan

EARTHY AND VEGETAL
- ○ flowers
- ○ grass
- ○ herbs
- ○ tobacco
- ○ tannins
- ○ leather
- ○ salt

- ○ medicine
- ○ metal
- ○ smoke

ALCOHOLIC
- ○ whiskey
- ○ sherry
- ○ brandy
- ○ cognac
- ○ wine

OTHER

STYLE
Choose One

- ○ Neat
- ○ With Water
- ○ On the Rocks
- ○ Cocktail

COLOR
Circle One

TASTING WHEEL
Rate your tasting experience, 1 (lowest) to 5 (highest)

TASTING NOTES

Describe the First Sip

....

....

....

Describe the Third Sip

....

....

....

Describe the Fade

....

....

....

What Is Most Striking About This Rum?

....

....

....

Additional Notes

....

....

....

....

GUIDED TASTING

Write It Out

QUALITY RATING   **COST RATING**   **OVERALL RATING**

RATE IT

Fill It In

ENJOYING RUM

RUM
The Details

Brand Name

Producer

Place of Origin

Price per Glass/Bottle

Distiller

Cask Type

Date

Age

ABV/Proof

Tasting Location

NOSE
Choose All That Apply

SWEET
- ○ caramel/toffee
- ○ brown sugar
- ○ molasses
- ○ honey
- ○ maple syrup
- ○ chocolate
- ○ cola
- ○ butterscotch
- ○ fruitcake
- ○ gingerbread

FRUITY
- ○ apple
- ○ apricot
- ○ cherry
- ○ currant
- ○ dark fruit
- ○ lemon
- ○ lime
- ○ orange
- ○ peach
- ○ banana

- ○ coconut
- ○ mango
- ○ pineapple

WOODY AND NUTTY
- ○ oak
- ○ cedar
- ○ sandalwood
- ○ coffee
- ○ almond
- ○ walnut

- ○ hazelnut
- ○ pecan

EARTHY AND VEGETAL
- ○ flowers
- ○ grass
- ○ herbs
- ○ tobacco
- ○ tannins
- ○ leather
- ○ salt

- ○ medicine
- ○ metal
- ○ smoke

ALCOHOLIC
- ○ whiskey
- ○ sherry
- ○ brandy
- ○ cognac
- ○ wine

OTHER

STYLE
Choose One

- ○ Neat
- ○ With Water
- ○ On the Rocks
- ○ Cocktail

TASTING WHEEL

Rate your tasting experience, 1 (lowest) to 5 (highest)

Legs · Sweetness · Molasses · Fruity · Woody · Nutty · Earthy · Vegetal · Spicy · Leathery · Smoky · Bite · Heat · Body · Balance · Linger

COLOR
Circle One

TASTING NOTES

Describe the First Sip

....

....

....

Describe the Third Sip

....

....

....

Describe the Fade

....

....

....

What Is Most Striking About This Rum?

....

....

....

Additional Notes

....

....

....

....

QUALITY RATING **COST RATING** **OVERALL RATING**

RUM

The Details

Brand Name
....

Producer | **Place of Origin** | **Price per Glass/Bottle**
....

Distiller | **Cask Type** | **Date**
....

Age | **ABV/Proof** | **Tasting Location**

NOSE

Choose All That Apply

SWEET
- caramel/toffee
- brown sugar
- molasses
- honey
- maple syrup
- chocolate
- cola
- butterscotch
- fruitcake
- gingerbread

FRUITY
- apple
- apricot
- cherry
- currant
- dark fruit
- lemon
- lime
- orange
- peach
- banana

- coconut
- mango
- pineapple

WOODY AND NUTTY
- oak
- cedar
- sandalwood
- coffee
- almond
- walnut

- hazelnut
- pecan

EARTHY AND VEGETAL
- flowers
- grass
- herbs
- tobacco
- tannins
- leather
- salt

- medicine
- metal
- smoke

ALCOHOLIC
- whiskey
- sherry
- brandy
- cognac
- wine

OTHER
....

STYLE

Choose One

- Neat
- With Water
- On the Rocks
- Cocktail

COLOR

Circle One

TASTING WHEEL

Rate your tasting experience, 1 (lowest) to 5 (highest)

TASTING NOTES

Describe the First Sip

....

....

....

Describe the Third Sip

....

....

....

Describe the Fade

....

....

....

What Is Most Striking About This Rum?

....

....

....

Additional Notes

....

....

....

....

GUIDED TASTING

Write It Out

QUALITY RATING   **COST RATING**   **OVERALL RATING**

RATE IT

Fill It In

RUM
The Details

Brand Name

Producer — Place of Origin — Price per Glass/Bottle

Distiller — Cask Type — Date

Age — ABV/Proof — Tasting Location

NOSE
Choose All That Apply

SWEET
- ○ caramel/toffee
- ○ brown sugar
- ○ molasses
- ○ honey
- ○ maple syrup
- ○ chocolate
- ○ cola
- ○ butterscotch
- ○ fruitcake
- ○ gingerbread

FRUITY
- ○ apple
- ○ apricot
- ○ cherry
- ○ currant
- ○ dark fruit
- ○ lemon
- ○ lime
- ○ orange
- ○ peach
- ○ banana

- ○ coconut
- ○ mango
- ○ pineapple

WOODY AND NUTTY
- ○ oak
- ○ cedar
- ○ sandalwood
- ○ coffee
- ○ almond
- ○ walnut

- ○ hazelnut
- ○ pecan

EARTHY AND VEGETAL
- ○ flowers
- ○ grass
- ○ herbs
- ○ tobacco
- ○ tannins
- ○ leather
- ○ salt

- ○ medicine
- ○ metal
- ○ smoke

ALCOHOLIC
- ○ whiskey
- ○ sherry
- ○ brandy
- ○ cognac
- ○ wine

OTHER

STYLE
Choose One

- ○ Neat
- ○ With Water
- ○ On the Rocks
- ○ Cocktail

COLOR
Circle One

TASTING WHEEL

Rate your tasting experience, 1 (lowest) to 5 (highest)

Legs, Sweetness, Molasses, Fruity, Woody, Nutty, Earthy, Vegetal, Spicy, Leathery, Smoky, Bite, Heat, Body, Balance, Linger

TASTING NOTES

Describe the First Sip

....
....
....

Describe the Third Sip

....
....
....

Describe the Fade

....
....
....

What Is Most Striking About This Rum?

....
....
....

Additional Notes

....
....
....
....

GUIDED TASTING

Write It Out

QUALITY RATING

COST RATING

OVERALL RATING

RATE IT

Fill It In

RUM

The Details

Brand Name

Producer   Place of Origin   Price per Glass/Bottle

Distiller   Cask Type   Date

Age   ABV/Proof   Tasting Location

NOSE

Choose All That Apply

SWEET
- ○ caramel/toffee
- ○ brown sugar
- ○ molasses
- ○ honey
- ○ maple syrup
- ○ chocolate
- ○ cola
- ○ butterscotch
- ○ fruitcake
- ○ gingerbread

FRUITY
- ○ apple
- ○ apricot
- ○ cherry
- ○ currant
- ○ dark fruit
- ○ lemon
- ○ lime
- ○ orange
- ○ peach
- ○ banana

- ○ coconut
- ○ mango
- ○ pineapple

WOODY AND NUTTY
- ○ oak
- ○ cedar
- ○ sandalwood
- ○ coffee
- ○ almond
- ○ walnut

- ○ hazelnut
- ○ pecan

EARTHY AND VEGETAL
- ○ flowers
- ○ grass
- ○ herbs
- ○ tobacco
- ○ tannins
- ○ leather
- ○ salt

- ○ medicine
- ○ metal
- ○ smoke

ALCOHOLIC
- ○ whiskey
- ○ sherry
- ○ brandy
- ○ cognac
- ○ wine

OTHER

STYLE

Choose One

- ○ Neat
- ○ With Water
- ○ On the Rocks
- ○ Cocktail

TASTING WHEEL

Rate your tasting experience, 1 (lowest) to 5 (highest)

COLOR

Circle One

TASTING NOTES

Describe the First Sip

....

....

....

Describe the Third Sip

....

....

....

Describe the Fade

....

....

....

What Is Most Striking About This Rum?

....

....

....

Additional Notes

....

....

....

....

GUIDED TASTING

Write It Out

QUALITY RATING　　　　**COST RATING**　　　　**OVERALL RATING**

☆☆☆☆☆　　　　

RATE IT

Fill It In

RUM
The Details

Brand Name

Producer   **Place of Origin**   **Price per Glass/Bottle**

Distiller   **Cask Type**   **Date**

Age   **ABV/Proof**   **Tasting Location**

NOSE
Choose All That Apply

SWEET
- ○ caramel/toffee
- ○ brown sugar
- ○ molasses
- ○ honey
- ○ maple syrup
- ○ chocolate
- ○ cola
- ○ butterscotch
- ○ fruitcake
- ○ gingerbread

FRUITY
- ○ apple
- ○ apricot
- ○ cherry
- ○ currant
- ○ dark fruit
- ○ lemon
- ○ lime
- ○ orange
- ○ peach
- ○ banana

- ○ coconut
- ○ mango
- ○ pineapple

WOODY AND NUTTY
- ○ oak
- ○ cedar
- ○ sandalwood
- ○ coffee
- ○ almond
- ○ walnut

- ○ hazelnut
- ○ pecan

EARTHY AND VEGETAL
- ○ flowers
- ○ grass
- ○ herbs
- ○ tobacco
- ○ tannins
- ○ leather
- ○ salt

- ○ medicine
- ○ metal
- ○ smoke

ALCOHOLIC
- ○ whiskey
- ○ sherry
- ○ brandy
- ○ cognac
- ○ wine

OTHER

STYLE
Choose One

- ○ Neat
- ○ With Water
- ○ On the Rocks
- ○ Cocktail

COLOR
Circle One

TASTING WHEEL

Rate your tasting experience, 1 (lowest) to 5 (highest)

Legs · Sweetness · Molasses · Fruity · Woody · Nutty · Earthy · Vegetal · Spicy · Leathery · Smoky · Bite · Heat · Body · Balance · Linger

TASTING NOTES

Describe the First Sip

....

....

....

Describe the Third Sip

....

....

....

Describe the Fade

....

....

....

What Is Most Striking About This Rum?

....

....

....

Additional Notes

....

....

....

....

GUIDED TASTING

Write It Out

QUALITY RATING

COST RATING

OVERALL RATING

RATE IT

Fill It In

RUM
The Details

Brand Name

Producer

Place of Origin

Price per Glass/Bottle

Distiller

Cask Type

Date

Age

ABV/Proof

Tasting Location

NOSE
Choose All That Apply

SWEET
- caramel/toffee
- brown sugar
- molasses
- honey
- maple syrup
- chocolate
- cola
- butterscotch
- fruitcake
- gingerbread

FRUITY
- apple
- apricot
- cherry
- currant
- dark fruit
- lemon
- lime
- orange
- peach
- banana

- coconut
- mango
- pineapple

WOODY AND NUTTY
- oak
- cedar
- sandalwood
- coffee
- almond
- walnut

- hazelnut
- pecan

EARTHY AND VEGETAL
- flowers
- grass
- herbs
- tobacco
- tannins
- leather
- salt

- medicine
- metal
- smoke

ALCOHOLIC
- whiskey
- sherry
- brandy
- cognac
- wine

OTHER

STYLE
Choose One

- Neat
- With Water
- On the Rocks
- Cocktail

TASTING WHEEL

Rate your tasting experience, 1 (lowest) to 5 (highest)

COLOR
Circle One

TASTING NOTES

Describe the First Sip

....

....

....

Describe the Third Sip

....

....

....

Describe the Fade

....

....

....

What Is Most Striking About This Rum?

....

....

....

Additional Notes

....

....

....

....

GUIDED TASTING

Write It Out

QUALITY RATING

COST RATING

OVERALL RATING

RATE IT

Fill It In

ENJOYING RUM

RUM
The Details

Brand Name

Producer   **Place of Origin**   **Price per Glass/Bottle**

Distiller   **Cask Type**   **Date**

Age   **ABV/Proof**   **Tasting Location**

NOSE
Choose All That Apply

SWEET
- caramel/toffee
- brown sugar
- molasses
- honey
- maple syrup
- chocolate
- cola
- butterscotch
- fruitcake
- gingerbread

FRUITY
- apple
- apricot
- cherry
- currant
- dark fruit
- lemon
- lime
- orange
- peach
- banana

- coconut
- mango
- pineapple

WOODY AND NUTTY
- oak
- cedar
- sandalwood
- coffee
- almond
- walnut

- hazelnut
- pecan

EARTHY AND VEGETAL
- flowers
- grass
- herbs
- tobacco
- tannins
- leather
- salt

- medicine
- metal
- smoke

ALCOHOLIC
- whiskey
- sherry
- brandy
- cognac
- wine

OTHER

STYLE
Choose One
- Neat
- With Water
- On the Rocks
- Cocktail

COLOR
Circle One

TASTING WHEEL

Rate your tasting experience, 1 (lowest) to 5 (highest)

Wheel spokes: Legs, Sweetness, Molasses, Fruity, Woody, Nutty, Earthy, Vegetal, Spicy, Leathery, Smoky, Bite, Heat, Body, Balance, Linger

TASTING NOTES

Describe the First Sip

Describe the Third Sip

Describe the Fade

What Is Most Striking About This Rum?

Additional Notes

GUIDED TASTING

Write It Out

QUALITY RATING

COST RATING

OVERALL RATING

RATE IT

Fill It In

RUM

The Details

Brand Name

Producer	**Place of Origin**	**Price per Glass/Bottle**
Distiller	**Cask Type**	**Date**
Age	**ABV/Proof**	**Tasting Location**

NOSE

Choose All That Apply

SWEET
- caramel/toffee
- brown sugar
- molasses
- honey
- maple syrup
- chocolate
- cola
- butterscotch
- fruitcake
- gingerbread

FRUITY
- apple
- apricot
- cherry
- currant
- dark fruit
- lemon
- lime
- orange
- peach
- banana

- coconut
- mango
- pineapple

WOODY AND NUTTY
- oak
- cedar
- sandalwood
- coffee
- almond
- walnut

- hazelnut
- pecan

EARTHY AND VEGETAL
- flowers
- grass
- herbs
- tobacco
- tannins
- leather
- salt

- medicine
- metal
- smoke

ALCOHOLIC
- whiskey
- sherry
- brandy
- cognac
- wine

OTHER

STYLE

Choose One

- Neat
- With Water
- On the Rocks
- Cocktail

TASTING WHEEL

Rate your tasting experience, 1 (lowest) to 5 (highest)

COLOR

Circle One

TASTING NOTES

Describe the First Sip

....

....

....

Describe the Third Sip

....

....

....

Describe the Fade

....

....

....

What Is Most Striking About This Rum?

....

....

....

Additional Notes

....

....

....

....

QUALITY RATING	COST RATING	OVERALL RATING

ENJOYING RUM

RUM
The Details

Brand Name

Producer   **Place of Origin**   **Price per Glass/Bottle**

Distiller   **Cask Type**   **Date**

Age   **ABV/Proof**   **Tasting Location**

NOSE
Choose All That Apply

SWEET
- ○ caramel/toffee
- ○ brown sugar
- ○ molasses
- ○ honey
- ○ maple syrup
- ○ chocolate
- ○ cola
- ○ butterscotch
- ○ fruitcake
- ○ gingerbread

FRUITY
- ○ apple
- ○ apricot
- ○ cherry
- ○ currant
- ○ dark fruit
- ○ lemon
- ○ lime
- ○ orange
- ○ peach
- ○ banana

- ○ coconut
- ○ mango
- ○ pineapple

WOODY AND NUTTY
- ○ oak
- ○ cedar
- ○ sandalwood
- ○ coffee
- ○ almond
- ○ walnut

- ○ hazelnut
- ○ pecan

EARTHY AND VEGETAL
- ○ flowers
- ○ grass
- ○ herbs
- ○ tobacco
- ○ tannins
- ○ leather
- ○ salt

- ○ medicine
- ○ metal
- ○ smoke

ALCOHOLIC
- ○ whiskey
- ○ sherry
- ○ brandy
- ○ cognac
- ○ wine

OTHER

STYLE
Choose One

- ○ Neat
- ○ With Water
- ○ On the Rocks
- ○ Cocktail

COLOR
Circle One

TASTING WHEEL

Rate your tasting experience, 1 (lowest) to 5 (highest)

Legs, Sweetness, Molasses, Fruity, Woody, Nutty, Earthy, Vegetal, Spicy, Leathery, Smoky, Bite, Heat, Body, Balance, Linger

Describe the First Sip

....

....

....

Describe the Third Sip

....

....

....

Describe the Fade

....

....

....

What Is Most Striking About This Rum?

....

....

....

Additional Notes

....

....

....

....

GUIDED TASTING · Write It Out

QUALITY RATING

☆ ☆ ☆ ☆ ☆

COST RATING

OVERALL RATING

RATE IT · Fill It In

RUM
The Details

Brand Name

Producer · Place of Origin · Price per Glass/Bottle

Distiller · Cask Type · Date

Age · ABV/Proof · Tasting Location

NOSE
Choose All That Apply

SWEET
- ○ caramel/toffee
- ○ brown sugar
- ○ molasses
- ○ honey
- ○ maple syrup
- ○ chocolate
- ○ cola
- ○ butterscotch
- ○ fruitcake
- ○ gingerbread

FRUITY
- ○ apple
- ○ apricot
- ○ cherry
- ○ currant
- ○ dark fruit
- ○ lemon
- ○ lime
- ○ orange
- ○ peach
- ○ banana

- ○ coconut
- ○ mango
- ○ pineapple

WOODY AND NUTTY
- ○ oak
- ○ cedar
- ○ sandalwood
- ○ coffee
- ○ almond
- ○ walnut

- ○ hazelnut
- ○ pecan

EARTHY AND VEGETAL
- ○ flowers
- ○ grass
- ○ herbs
- ○ tobacco
- ○ tannins
- ○ leather
- ○ salt

- ○ medicine
- ○ metal
- ○ smoke

ALCOHOLIC
- ○ whiskey
- ○ sherry
- ○ brandy
- ○ cognac
- ○ wine

OTHER

STYLE
Choose One

- ○ Neat
- ○ With Water
- ○ On the Rocks
- ○ Cocktail

COLOR
Circle One

TASTING WHEEL

Rate your tasting experience, 1 (lowest) to 5 (highest)

TASTING NOTES

Describe the First Sip

....

....

....

Describe the Third Sip

....

....

....

Describe the Fade

....

....

....

What Is Most Striking About This Rum?

....

....

....

Additional Notes

....

....

....

....

QUALITY RATING   **COST RATING**   **OVERALL RATING**

RATE IT

Fill It In

RUM
The Details

Brand Name

Producer

Place of Origin

Price per Glass/Bottle

Distiller

Cask Type

Date

Age

ABV/Proof

Tasting Location

NOSE
Choose All That Apply

SWEET
- ○ caramel/toffee
- ○ brown sugar
- ○ molasses
- ○ honey
- ○ maple syrup
- ○ chocolate
- ○ cola
- ○ butterscotch
- ○ fruitcake
- ○ gingerbread

FRUITY
- ○ apple
- ○ apricot
- ○ cherry
- ○ currant
- ○ dark fruit
- ○ lemon
- ○ lime
- ○ orange
- ○ peach
- ○ banana

- ○ coconut
- ○ mango
- ○ pineapple

WOODY AND NUTTY
- ○ oak
- ○ cedar
- ○ sandalwood
- ○ coffee
- ○ almond
- ○ walnut

- ○ hazelnut
- ○ pecan

EARTHY AND VEGETAL
- ○ flowers
- ○ grass
- ○ herbs
- ○ tobacco
- ○ tannins
- ○ leather
- ○ salt

- ○ medicine
- ○ metal
- ○ smoke

ALCOHOLIC
- ○ whiskey
- ○ sherry
- ○ brandy
- ○ cognac
- ○ wine

OTHER

STYLE
Choose One

- ○ Neat
- ○ With Water
- ○ On the Rocks
- ○ Cocktail

COLOR
Circle One

TASTING WHEEL

Rate your tasting experience, 1 (lowest) to 5 (highest)

Legs · Sweetness · Molasses · Fruity · Woody · Nutty · Earthy · Vegetal · Spicy · Leathery · Smoky · Bite · Heat · Body · Balance · Linger

Describe the First Sip

....

....

....

Describe the Third Sip

....

....

....

Describe the Fade

....

....

....

What Is Most Striking About This Rum?

....

....

....

Additional Notes

....

....

....

....

GUIDED TASTING

Write It Out

QUALITY RATING

COST RATING

OVERALL RATING

RATE IT

Fill It In

RUM

The Details

Brand Name

Producer | Place of Origin | Price per Glass/Bottle

Distiller | Cask Type | Date

Age | ABV/Proof | Tasting Location

NOSE

Choose All That Apply

SWEET
- ○ caramel/toffee
- ○ brown sugar
- ○ molasses
- ○ honey
- ○ maple syrup
- ○ chocolate
- ○ cola
- ○ butterscotch
- ○ fruitcake
- ○ gingerbread

FRUITY
- ○ apple
- ○ apricot
- ○ cherry
- ○ currant
- ○ dark fruit
- ○ lemon
- ○ lime
- ○ orange
- ○ peach
- ○ banana

- ○ coconut
- ○ mango
- ○ pineapple

WOODY AND NUTTY
- ○ oak
- ○ cedar
- ○ sandalwood
- ○ coffee
- ○ almond
- ○ walnut

- ○ hazelnut
- ○ pecan

EARTHY AND VEGETAL
- ○ flowers
- ○ grass
- ○ herbs
- ○ tobacco
- ○ tannins
- ○ leather
- ○ salt

- ○ medicine
- ○ metal
- ○ smoke

ALCOHOLIC
- ○ whiskey
- ○ sherry
- ○ brandy
- ○ cognac
- ○ wine

OTHER

STYLE

Choose One

- ○ Neat
- ○ With Water
- ○ On the Rocks
- ○ Cocktail

COLOR

Circle One

TASTING WHEEL

Rate your tasting experience, 1 (lowest) to 5 (highest)

Legs · Linger · Sweetness · Balance · Molasses · Body · Fruity · Heat · Woody · Bite · Nutty · Smoky · Earthy · Leathery · Vegetal · Spicy

TASTING NOTES

Describe the First Sip

....

....

....

Describe the Third Sip

....

....

....

Describe the Fade

....

....

....

What Is Most Striking About This Rum?

....

....

....

Additional Notes

....

....

....

....

QUALITY RATING	COST RATING	OVERALL RATING

RUM
The Details

Brand Name

Producer   **Place of Origin**   **Price per Glass/Bottle**

Distiller   **Cask Type**   **Date**

Age   **ABV/Proof**   **Tasting Location**

NOSE
Choose All That Apply

SWEET
- caramel/toffee
- brown sugar
- molasses
- honey
- maple syrup
- chocolate
- cola
- butterscotch
- fruitcake
- gingerbread

FRUITY
- apple
- apricot
- cherry
- currant
- dark fruit
- lemon
- lime
- orange
- peach
- banana

- coconut
- mango
- pineapple

WOODY AND NUTTY
- oak
- cedar
- sandalwood
- coffee
- almond
- walnut

- hazelnut
- pecan

EARTHY AND VEGETAL
- flowers
- grass
- herbs
- tobacco
- tannins
- leather
- salt

- medicine
- metal
- smoke

ALCOHOLIC
- whiskey
- sherry
- brandy
- cognac
- wine

OTHER

STYLE
Choose One

- Neat
- With Water
- On the Rocks
- Cocktail

COLOR
Circle One

TASTING WHEEL
Rate your tasting experience, 1 (lowest) to 5 (highest)

Legs, Sweetness, Molasses, Fruity, Woody, Nutty, Earthy, Vegetal, Spicy, Leathery, Smoky, Bite, Heat, Body, Balance, Linger

TASTING NOTES

Describe the First Sip

....

....

....

Describe the Third Sip

....

....

....

Describe the Fade

....

....

....

What Is Most Striking About This Rum?

....

....

....

Additional Notes

....

....

....

....

GUIDED TASTING

Write It Out

QUALITY RATING	COST RATING	OVERALL RATING

RATE IT

Fill It In

ENJOYING RUM

RUM — The Details

Brand Name

Producer   Place of Origin   Price per Glass/Bottle

Distiller   Cask Type   Date

Age   ABV/Proof   Tasting Location

NOSE — Choose All That Apply

SWEET
- caramel/toffee
- brown sugar
- molasses
- honey
- maple syrup
- chocolate
- cola
- butterscotch
- fruitcake
- gingerbread

FRUITY
- apple
- apricot
- cherry
- currant
- dark fruit
- lemon
- lime
- orange
- peach
- banana

- coconut
- mango
- pineapple

WOODY AND NUTTY
- oak
- cedar
- sandalwood
- coffee
- almond
- walnut

- hazelnut
- pecan

EARTHY AND VEGETAL
- flowers
- grass
- herbs
- tobacco
- tannins
- leather
- salt

- medicine
- metal
- smoke

ALCOHOLIC
- whiskey
- sherry
- brandy
- cognac
- wine

OTHER

STYLE — Choose One

- Neat
- With Water
- On the Rocks
- Cocktail

COLOR — Circle One

TASTING WHEEL

Rate your tasting experience, 1 (lowest) to 5 (highest)

Legs, Sweetness, Molasses, Fruity, Woody, Nutty, Earthy, Vegetal, Spicy, Leathery, Smoky, Bite, Heat, Body, Balance, Linger

Describe the First Sip

....

....

....

Describe the Third Sip

....

....

....

Describe the Fade

....

....

....

What Is Most Striking About This Rum?

....

....

....

Additional Notes

....

....

....

....

GUIDED TASTING

Write It Out

QUALITY RATING

COST RATING

OVERALL RATING

RATE IT

Fill It In

ENJOYING RUM

RUM
The Details

Brand Name

Producer | Place of Origin | Price per Glass/Bottle

Distiller | Cask Type | Date

Age | ABV/Proof | Tasting Location

NOSE
Choose All That Apply

SWEET
- ○ caramel/toffee
- ○ brown sugar
- ○ molasses
- ○ honey
- ○ maple syrup
- ○ chocolate
- ○ cola
- ○ butterscotch
- ○ fruitcake
- ○ gingerbread

FRUITY
- ○ apple
- ○ apricot
- ○ cherry
- ○ currant
- ○ dark fruit
- ○ lemon
- ○ lime
- ○ orange
- ○ peach
- ○ banana

- ○ coconut
- ○ mango
- ○ pineapple

WOODY AND NUTTY
- ○ oak
- ○ cedar
- ○ sandalwood
- ○ coffee
- ○ almond
- ○ walnut

- ○ hazelnut
- ○ pecan

EARTHY AND VEGETAL
- ○ flowers
- ○ grass
- ○ herbs
- ○ tobacco
- ○ tannins
- ○ leather
- ○ salt

- ○ medicine
- ○ metal
- ○ smoke

ALCOHOLIC
- ○ whiskey
- ○ sherry
- ○ brandy
- ○ cognac
- ○ wine

OTHER

STYLE
Choose One

- ○ Neat
- ○ With Water
- ○ On the Rocks
- ○ Cocktail

COLOR
Circle One

TASTING WHEEL

Rate your tasting experience, 1 (lowest) to 5 (highest)

Legs, Sweetness, Molasses, Fruity, Woody, Nutty, Earthy, Vegetal, Spicy, Leathery, Smoky, Bite, Heat, Body, Balance, Linger

TASTING NOTES

Describe the First Sip

....

....

....

Describe the Third Sip

....

....

....

Describe the Fade

....

....

....

What Is Most Striking About This Rum?

....

....

....

Additional Notes

....

....

....

....

QUALITY RATING	COST RATING	OVERALL RATING

RUM
The Details

Brand Name

Producer

Place of Origin

Price per Glass/Bottle

Distiller

Cask Type

Date

Age

ABV/Proof

Tasting Location

NOSE
Choose All That Apply

SWEET
- ○ caramel/toffee
- ○ brown sugar
- ○ molasses
- ○ honey
- ○ maple syrup
- ○ chocolate
- ○ cola
- ○ butterscotch
- ○ fruitcake
- ○ gingerbread

FRUITY
- ○ apple
- ○ apricot
- ○ cherry
- ○ currant
- ○ dark fruit
- ○ lemon
- ○ lime
- ○ orange
- ○ peach
- ○ banana

- ○ coconut
- ○ mango
- ○ pineapple

WOODY AND NUTTY
- ○ oak
- ○ cedar
- ○ sandalwood
- ○ coffee
- ○ almond
- ○ walnut

- ○ hazelnut
- ○ pecan

EARTHY AND VEGETAL
- ○ flowers
- ○ grass
- ○ herbs
- ○ tobacco
- ○ tannins
- ○ leather
- ○ salt

- ○ medicine
- ○ metal
- ○ smoke

ALCOHOLIC
- ○ whiskey
- ○ sherry
- ○ brandy
- ○ cognac
- ○ wine

OTHER

STYLE
Choose One

- ○ Neat
- ○ With Water
- ○ On the Rocks
- ○ Cocktail

COLOR
Circle One

TASTING WHEEL

Rate your tasting experience, 1 (lowest) to 5 (highest)

Legs · Sweetness · Molasses · Fruity · Woody · Nutty · Earthy · Vegetal · Spicy · Leathery · Smoky · Bite · Heat · Body · Balance · Linger

TASTING NOTES

Describe the First Sip

....

....

....

Describe the Third Sip

....

....

....

Describe the Fade

....

....

....

What Is Most Striking About This Rum?

....

....

....

Additional Notes

....

....

....

....

QUALITY RATING

COST RATING

OVERALL RATING

RUM
The Details

Brand Name

Producer

Place of Origin

Price per Glass/Bottle

Distiller

Cask Type

Date

Age

ABV/Proof

Tasting Location

NOSE
Choose All That Apply

SWEET
- ○ caramel/toffee
- ○ brown sugar
- ○ molasses
- ○ honey
- ○ maple syrup
- ○ chocolate
- ○ cola
- ○ butterscotch
- ○ fruitcake
- ○ gingerbread

FRUITY
- ○ apple
- ○ apricot
- ○ cherry
- ○ currant
- ○ dark fruit
- ○ lemon
- ○ lime
- ○ orange
- ○ peach
- ○ banana

- ○ coconut
- ○ mango
- ○ pineapple

WOODY AND NUTTY
- ○ oak
- ○ cedar
- ○ sandalwood
- ○ coffee
- ○ almond
- ○ walnut

- ○ hazelnut
- ○ pecan

EARTHY AND VEGETAL
- ○ flowers
- ○ grass
- ○ herbs
- ○ tobacco
- ○ tannins
- ○ leather
- ○ salt

- ○ medicine
- ○ metal
- ○ smoke

ALCOHOLIC
- ○ whiskey
- ○ sherry
- ○ brandy
- ○ cognac
- ○ wine

OTHER

STYLE
Choose One

- ○ Neat
- ○ With Water
- ○ On the Rocks
- ○ Cocktail

COLOR
Circle One

TASTING WHEEL

Rate your tasting experience, 1 (lowest) to 5 (highest)

Legs · Sweetness · Molasses · Fruity · Woody · Nutty · Earthy · Vegetal · Spicy · Leathery · Smoky · Bite · Heat · Body · Balance · Linger

Describe the First Sip

....

....

....

Describe the Third Sip

....

....

....

Describe the Fade

....

....

....

What Is Most Striking About This Rum?

....

....

....

Additional Notes

....

....

....

....

GUIDED TASTING

Write It Out

QUALITY RATING

COST RATING

OVERALL RATING

RATE IT

Fill It In

RUM
The Details

Brand Name

Producer   Place of Origin   Price per Glass/Bottle

Distiller   Cask Type   Date

Age   ABV/Proof   Tasting Location

NOSE
Choose All That Apply

SWEET
- ○ caramel/toffee
- ○ brown sugar
- ○ molasses
- ○ honey
- ○ maple syrup
- ○ chocolate
- ○ cola
- ○ butterscotch
- ○ fruitcake
- ○ gingerbread

FRUITY
- ○ apple
- ○ apricot
- ○ cherry
- ○ currant
- ○ dark fruit
- ○ lemon
- ○ lime
- ○ orange
- ○ peach
- ○ banana

- ○ coconut
- ○ mango
- ○ pineapple

WOODY AND NUTTY
- ○ oak
- ○ cedar
- ○ sandalwood
- ○ coffee
- ○ almond
- ○ walnut

- ○ hazelnut
- ○ pecan

EARTHY AND VEGETAL
- ○ flowers
- ○ grass
- ○ herbs
- ○ tobacco
- ○ tannins
- ○ leather
- ○ salt

- ○ medicine
- ○ metal
- ○ smoke

ALCOHOLIC
- ○ whiskey
- ○ sherry
- ○ brandy
- ○ cognac
- ○ wine

OTHER
....

STYLE
Choose One

- ○ Neat
- ○ With Water
- ○ On the Rocks
- ○ Cocktail

COLOR
Circle One

TASTING WHEEL

Rate your tasting experience, 1 (lowest) to 5 (highest)

TASTING NOTES

Describe the First Sip

....

....

....

Describe the Third Sip

....

....

....

Describe the Fade

....

....

....

What Is Most Striking About This Rum?

....

....

....

Additional Notes

....

....

....

....

GUIDED TASTING

Write It Out

QUALITY RATING   **COST RATING**   **OVERALL RATING**

RATE IT

Fill It In

RUM
The Details

Brand Name

Producer | Place of Origin | Price per Glass/Bottle

Distiller | Cask Type | Date

Age | ABV/Proof | Tasting Location

NOSE
Choose All That Apply

SWEET
- ○ caramel/toffee
- ○ brown sugar
- ○ molasses
- ○ honey
- ○ maple syrup
- ○ chocolate
- ○ cola
- ○ butterscotch
- ○ fruitcake
- ○ gingerbread

FRUITY
- ○ apple
- ○ apricot
- ○ cherry
- ○ currant
- ○ dark fruit
- ○ lemon
- ○ lime
- ○ orange
- ○ peach
- ○ banana

- ○ coconut
- ○ mango
- ○ pineapple

WOODY AND NUTTY
- ○ oak
- ○ cedar
- ○ sandalwood
- ○ coffee
- ○ almond
- ○ walnut

- ○ hazelnut
- ○ pecan

EARTHY AND VEGETAL
- ○ flowers
- ○ grass
- ○ herbs
- ○ tobacco
- ○ tannins
- ○ leather
- ○ salt

- ○ medicine
- ○ metal
- ○ smoke

ALCOHOLIC
- ○ whiskey
- ○ sherry
- ○ brandy
- ○ cognac
- ○ wine

OTHER

STYLE
Choose One

- ○ Neat
- ○ With Water
- ○ On the Rocks
- ○ Cocktail

COLOR
Circle One

TASTING WHEEL

Rate your tasting experience, 1 (lowest) to 5 (highest)

TASTING NOTES

Describe the First Sip

....

....

....

Describe the Third Sip

....

....

....

Describe the Fade

....

....

....

What Is Most Striking About This Rum?

....

....

....

Additional Notes

....

....

....

....

GUIDED TASTING

Write It Out

QUALITY RATING

COST RATING

OVERALL RATING

RATE IT

Fill It In

RUM
The Details

Brand Name ....

Producer .... | Place of Origin .... | Price per Glass/Bottle

Distiller .... | Cask Type .... | Date

Age .... | ABV/Proof .... | Tasting Location

NOSE
Choose All That Apply

SWEET
- ◯ caramel/toffee
- ◯ brown sugar
- ◯ molasses
- ◯ honey
- ◯ maple syrup
- ◯ chocolate
- ◯ cola
- ◯ butterscotch
- ◯ fruitcake
- ◯ gingerbread

FRUITY
- ◯ apple
- ◯ apricot
- ◯ cherry
- ◯ currant
- ◯ dark fruit
- ◯ lemon
- ◯ lime
- ◯ orange
- ◯ peach
- ◯ banana

- ◯ coconut
- ◯ mango
- ◯ pineapple

WOODY AND NUTTY
- ◯ oak
- ◯ cedar
- ◯ sandalwood
- ◯ coffee
- ◯ almond
- ◯ walnut

- ◯ hazelnut
- ◯ pecan

EARTHY AND VEGETAL
- ◯ flowers
- ◯ grass
- ◯ herbs
- ◯ tobacco
- ◯ tannins
- ◯ leather
- ◯ salt

- ◯ medicine
- ◯ metal
- ◯ smoke

ALCOHOLIC
- ◯ whiskey
- ◯ sherry
- ◯ brandy
- ◯ cognac
- ◯ wine

OTHER
....

STYLE
Choose One

- ◯ Neat
- ◯ With Water
- ◯ On the Rocks
- ◯ Cocktail

TASTING WHEEL

Rate your tasting experience, 1 (lowest) to 5 (highest)

Legs, Linger, Sweetness, Balance, Molasses, Body, Fruity, Heat, Woody, Bite, Nutty, Smoky, Earthy, Leathery, Spicy, Vegetal

COLOR
Circle One

TASTING NOTES

Describe the First Sip

....

....

....

Describe the Third Sip

....

....

....

Describe the Fade

....

....

....

What Is Most Striking About This Rum?

....

....

....

Additional Notes

....

....

....

....

GUIDED TASTING

Write It Out

QUALITY RATING	COST RATING	OVERALL RATING
☆☆☆☆☆		🍾🍾🍾🍾🍾

RATE IT

Fill It In

RUM

The Details

Brand Name

Producer   **Place of Origin**   **Price per Glass/Bottle**

Distiller   **Cask Type**   **Date**

Age   **ABV/Proof**   **Tasting Location**

NOSE

Choose All That Apply

SWEET
- ○ caramel/toffee
- ○ brown sugar
- ○ molasses
- ○ honey
- ○ maple syrup
- ○ chocolate
- ○ cola
- ○ butterscotch
- ○ fruitcake
- ○ gingerbread

FRUITY
- ○ apple
- ○ apricot
- ○ cherry
- ○ currant
- ○ dark fruit
- ○ lemon
- ○ lime
- ○ orange
- ○ peach
- ○ banana

- ○ coconut
- ○ mango
- ○ pineapple

WOODY AND NUTTY
- ○ oak
- ○ cedar
- ○ sandalwood
- ○ coffee
- ○ almond
- ○ walnut

- ○ hazelnut
- ○ pecan

EARTHY AND VEGETAL
- ○ flowers
- ○ grass
- ○ herbs
- ○ tobacco
- ○ tannins
- ○ leather
- ○ salt

- ○ medicine
- ○ metal
- ○ smoke

ALCOHOLIC
- ○ whiskey
- ○ sherry
- ○ brandy
- ○ cognac
- ○ wine

OTHER

STYLE

Choose One

- ○ Neat
- ○ With Water
- ○ On the Rocks
- ○ Cocktail

TASTING WHEEL

Rate your tasting experience, 1 (lowest) to 5 (highest)

Legs, Sweetness, Molasses, Fruity, Woody, Nutty, Earthy, Vegetal, Spicy, Leathery, Smoky, Bite, Heat, Body, Balance, Linger

COLOR

Circle One

Describe the First Sip

....

....

....

Describe the Third Sip

....

....

....

Describe the Fade

....

....

....

What Is Most Striking About This Rum?

....

....

....

Additional Notes

....

....

....

....

....

GUIDED TASTING

Write It Out

QUALITY RATING	COST RATING	OVERALL RATING

RATE IT

Fill It In

ENJOYING RUM

RUM
The Details

Brand Name

Producer | **Place of Origin** | **Price per Glass/Bottle**

Distiller | **Cask Type** | **Date**

Age | **ABV/Proof** | **Tasting Location**

NOSE
Choose All That Apply

SWEET
- ○ caramel/toffee
- ○ brown sugar
- ○ molasses
- ○ honey
- ○ maple syrup
- ○ chocolate
- ○ cola
- ○ butterscotch
- ○ fruitcake
- ○ gingerbread

FRUITY
- ○ apple
- ○ apricot
- ○ cherry
- ○ currant
- ○ dark fruit
- ○ lemon
- ○ lime
- ○ orange
- ○ peach
- ○ banana

- ○ coconut
- ○ mango
- ○ pineapple

WOODY AND NUTTY
- ○ oak
- ○ cedar
- ○ sandalwood
- ○ coffee
- ○ almond
- ○ walnut

- ○ hazelnut
- ○ pecan

EARTHY AND VEGETAL
- ○ flowers
- ○ grass
- ○ herbs
- ○ tobacco
- ○ tannins
- ○ leather
- ○ salt

- ○ medicine
- ○ metal
- ○ smoke

ALCOHOLIC
- ○ whiskey
- ○ sherry
- ○ brandy
- ○ cognac
- ○ wine

OTHER

STYLE
Choose One

- ○ Neat
- ○ With Water
- ○ On the Rocks
- ○ Cocktail

COLOR
Circle One

TASTING WHEEL

Rate your tasting experience, 1 (lowest) to 5 (highest)

Legs, Sweetness, Molasses, Fruity, Woody, Nutty, Earthy, Vegetal, Spicy, Leathery, Smoky, Bite, Heat, Body, Balance, Linger

TASTING NOTES

Describe the First Sip

....

....

....

Describe the Third Sip

....

....

....

Describe the Fade

....

....

....

What Is Most Striking About This Rum?

....

....

....

Additional Notes

....

....

....

....

GUIDED TASTING

Write It Out

QUALITY RATING

COST RATING

OVERALL RATING

RATE IT

Fill It In

RUM
The Details

Brand Name

Producer

Place of Origin

Price per Glass/Bottle

Distiller

Cask Type

Date

Age

ABV/Proof

Tasting Location

NOSE
Choose All That Apply

SWEET
- ○ caramel/toffee
- ○ brown sugar
- ○ molasses
- ○ honey
- ○ maple syrup
- ○ chocolate
- ○ cola
- ○ butterscotch
- ○ fruitcake
- ○ gingerbread

FRUITY
- ○ apple
- ○ apricot
- ○ cherry
- ○ currant
- ○ dark fruit
- ○ lemon
- ○ lime
- ○ orange
- ○ peach
- ○ banana

- ○ coconut
- ○ mango
- ○ pineapple

WOODY AND NUTTY
- ○ oak
- ○ cedar
- ○ sandalwood
- ○ coffee
- ○ almond
- ○ walnut

- ○ hazelnut
- ○ pecan

EARTHY AND VEGETAL
- ○ flowers
- ○ grass
- ○ herbs
- ○ tobacco
- ○ tannins
- ○ leather
- ○ salt

- ○ medicine
- ○ metal
- ○ smoke

ALCOHOLIC
- ○ whiskey
- ○ sherry
- ○ brandy
- ○ cognac
- ○ wine

OTHER

STYLE
Choose One

- ○ Neat
- ○ With Water
- ○ On the Rocks
- ○ Cocktail

COLOR
Circle One

TASTING WHEEL

Rate your tasting experience, 1 (lowest) to 5 (highest)

Legs, Linger, Sweetness, Balance, Molasses, Body, Fruity, Heat, Woody, Bite, Nutty, Smoky, Earthy, Leathery, Spicy, Vegetal

TASTING NOTES

Describe the First Sip

....

....

....

Describe the Third Sip

....

....

....

Describe the Fade

....

....

....

What Is Most Striking About This Rum?

....

....

....

Additional Notes

....

....

....

....

GUIDED TASTING

Write It Out

QUALITY RATING   COST RATING   OVERALL RATING

RATE IT

Fill It In

ENJOYING RUM

RUM — The Details

Brand Name

Producer   **Place of Origin**   **Price per Glass/Bottle**

Distiller   **Cask Type**   **Date**

Age   **ABV/Proof**   **Tasting Location**

NOSE — Choose All That Apply

SWEET
- ○ caramel/toffee
- ○ brown sugar
- ○ molasses
- ○ honey
- ○ maple syrup
- ○ chocolate
- ○ cola
- ○ butterscotch
- ○ fruitcake
- ○ gingerbread

FRUITY
- ○ apple
- ○ apricot
- ○ cherry
- ○ currant
- ○ dark fruit
- ○ lemon
- ○ lime
- ○ orange
- ○ peach
- ○ banana

- ○ coconut
- ○ mango
- ○ pineapple

WOODY AND NUTTY
- ○ oak
- ○ cedar
- ○ sandalwood
- ○ coffee
- ○ almond
- ○ walnut

- ○ hazelnut
- ○ pecan

EARTHY AND VEGETAL
- ○ flowers
- ○ grass
- ○ herbs
- ○ tobacco
- ○ tannins
- ○ leather
- ○ salt

- ○ medicine
- ○ metal
- ○ smoke

ALCOHOLIC
- ○ whiskey
- ○ sherry
- ○ brandy
- ○ cognac
- ○ wine

OTHER

STYLE — Choose One

- ○ Neat
- ○ With Water
- ○ On the Rocks
- ○ Cocktail

COLOR — Circle One

TASTING WHEEL

Rate your tasting experience, 1 (lowest) to 5 (highest)

Legs, Sweetness, Molasses, Fruity, Woody, Nutty, Earthy, Vegetal, Spicy, Leathery, Smoky, Bite, Heat, Body, Balance, Linger

TASTING NOTES

Describe the First Sip

....

....

....

Describe the Third Sip

....

....

....

Describe the Fade

....

....

....

What Is Most Striking About This Rum?

....

....

....

Additional Notes

....

....

....

....

QUALITY RATING	COST RATING	OVERALL RATING

RUM
The Details

Brand Name

Producer   Place of Origin   Price per Glass/Bottle

Distiller   Cask Type   Date

Age   ABV/Proof   Tasting Location

NOSE
Choose All That Apply

SWEET
- ○ caramel/toffee
- ○ brown sugar
- ○ molasses
- ○ honey
- ○ maple syrup
- ○ chocolate
- ○ cola
- ○ butterscotch
- ○ fruitcake
- ○ gingerbread

FRUITY
- ○ apple
- ○ apricot
- ○ cherry
- ○ currant
- ○ dark fruit
- ○ lemon
- ○ lime
- ○ orange
- ○ peach
- ○ banana

- ○ coconut
- ○ mango
- ○ pineapple

WOODY AND NUTTY
- ○ oak
- ○ cedar
- ○ sandalwood
- ○ coffee
- ○ almond
- ○ walnut

- ○ hazelnut
- ○ pecan

EARTHY AND VEGETAL
- ○ flowers
- ○ grass
- ○ herbs
- ○ tobacco
- ○ tannins
- ○ leather
- ○ salt

- ○ medicine
- ○ metal
- ○ smoke

ALCOHOLIC
- ○ whiskey
- ○ sherry
- ○ brandy
- ○ cognac
- ○ wine

OTHER

STYLE
Choose One

- ○ Neat
- ○ With Water
- ○ On the Rocks
- ○ Cocktail

COLOR
Circle One

TASTING WHEEL

Rate your tasting experience, 1 (lowest) to 5 (highest)

Describe the First Sip

Describe the Third Sip

Describe the Fade

What Is Most Striking About This Rum?

Additional Notes

QUALITY RATING

COST RATING

OVERALL RATING

RUM
The Details

Brand Name

Producer   **Place of Origin**   **Price per Glass/Bottle**

Distiller   **Cask Type**   **Date**

Age   **ABV/Proof**   **Tasting Location**

NOSE
Choose All That Apply

SWEET
- ○ caramel/toffee
- ○ brown sugar
- ○ molasses
- ○ honey
- ○ maple syrup
- ○ chocolate
- ○ cola
- ○ butterscotch
- ○ fruitcake
- ○ gingerbread

FRUITY
- ○ apple
- ○ apricot
- ○ cherry
- ○ currant
- ○ dark fruit
- ○ lemon
- ○ lime
- ○ orange
- ○ peach
- ○ banana

- ○ coconut
- ○ mango
- ○ pineapple

WOODY AND NUTTY
- ○ oak
- ○ cedar
- ○ sandalwood
- ○ coffee
- ○ almond
- ○ walnut

- ○ hazelnut
- ○ pecan

EARTHY AND VEGETAL
- ○ flowers
- ○ grass
- ○ herbs
- ○ tobacco
- ○ tannins
- ○ leather
- ○ salt

- ○ medicine
- ○ metal
- ○ smoke

ALCOHOLIC
- ○ whiskey
- ○ sherry
- ○ brandy
- ○ cognac
- ○ wine

OTHER

STYLE
Choose One

- ○ Neat
- ○ With Water
- ○ On the Rocks
- ○ Cocktail

COLOR
Circle One

TASTING WHEEL

Rate your tasting experience, 1 (lowest) to 5 (highest)

Legs · Linger · Balance · Body · Heat · Bite · Smoky · Leathery · Spicy · Vegetal · Earthy · Nutty · Woody · Fruity · Molasses · Sweetness

TASTING NOTES

Describe the First Sip

....

....

....

Describe the Third Sip

....

....

....

Describe the Fade

....

....

....

What Is Most Striking About This Rum?

....

....

....

Additional Notes

....

....

....

....

GUIDED TASTING

Write It Out

QUALITY RATING	COST RATING	OVERALL RATING

RATE IT

Fill It In

RUM
The Details

Brand Name

Producer   Place of Origin   Price per Glass/Bottle

Distiller   Cask Type   Date

Age   ABV/Proof   Tasting Location

NOSE
Choose All That Apply

SWEET
- caramel/toffee
- brown sugar
- molasses
- honey
- maple syrup
- chocolate
- cola
- butterscotch
- fruitcake
- gingerbread

FRUITY
- apple
- apricot
- cherry
- currant
- dark fruit
- lemon
- lime
- orange
- peach
- banana

- coconut
- mango
- pineapple

WOODY AND NUTTY
- oak
- cedar
- sandalwood
- coffee
- almond
- walnut

- hazelnut
- pecan

EARTHY AND VEGETAL
- flowers
- grass
- herbs
- tobacco
- tannins
- leather
- salt

- medicine
- metal
- smoke

ALCOHOLIC
- whiskey
- sherry
- brandy
- cognac
- wine

OTHER

STYLE
Choose One

- Neat
- With Water
- On the Rocks
- Cocktail

COLOR
Circle One

TASTING WHEEL

Rate your tasting experience, 1 (lowest) to 5 (highest)

Legs, Sweetness, Molasses, Fruity, Woody, Nutty, Earthy, Vegetal, Spicy, Leathery, Smoky, Bite, Heat, Body, Balance, Linger

TASTING NOTES

Describe the First Sip

....

....

....

Describe the Third Sip

....

....

....

Describe the Fade

....

....

....

What Is Most Striking About This Rum?

....

....

....

Additional Notes

....

....

....

....

GUIDED TASTING

Write It Out

QUALITY RATING

COST RATING

OVERALL RATING

RATE IT

Fill It In

RUM
The Details

Brand Name

Producer

Place of Origin

Price per Glass/Bottle

Distiller

Cask Type

Date

Age

ABV/Proof

Tasting Location

NOSE
Choose All That Apply

SWEET
- ◯ caramel/toffee
- ◯ brown sugar
- ◯ molasses
- ◯ honey
- ◯ maple syrup
- ◯ chocolate
- ◯ cola
- ◯ butterscotch
- ◯ fruitcake
- ◯ gingerbread

FRUITY
- ◯ apple
- ◯ apricot
- ◯ cherry
- ◯ currant
- ◯ dark fruit
- ◯ lemon
- ◯ lime
- ◯ orange
- ◯ peach
- ◯ banana

- ◯ coconut
- ◯ mango
- ◯ pineapple

WOODY AND NUTTY
- ◯ oak
- ◯ cedar
- ◯ sandalwood
- ◯ coffee
- ◯ almond
- ◯ walnut

- ◯ hazelnut
- ◯ pecan

EARTHY AND VEGETAL
- ◯ flowers
- ◯ grass
- ◯ herbs
- ◯ tobacco
- ◯ tannins
- ◯ leather
- ◯ salt

- ◯ medicine
- ◯ metal
- ◯ smoke

ALCOHOLIC
- ◯ whiskey
- ◯ sherry
- ◯ brandy
- ◯ cognac
- ◯ wine

OTHER

STYLE
Choose One

- ◯ Neat
- ◯ With Water
- ◯ On the Rocks
- ◯ Cocktail

COLOR
Circle One

TASTING WHEEL

Rate your tasting experience, 1 (lowest) to 5 (highest)

Legs, Linger, Balance, Body, Heat, Bite, Smoky, Leathery, Spicy, Vegetal, Earthy, Nutty, Woody, Fruity, Molasses, Sweetness

TASTING NOTES

Describe the First Sip

....
....
....

Describe the Third Sip

....
....
....

Describe the Fade

....
....
....

What Is Most Striking About This Rum?

....
....
....

Additional Notes

....
....
....
....

QUALITY RATING

COST RATING

OVERALL RATING

RATE IT

Fill It In

RUM
The Details

Brand Name

Producer   Place of Origin   Price per Glass/Bottle

Distiller   Cask Type   Date

Age   ABV/Proof   Tasting Location

NOSE
Choose All That Apply

SWEET
- ○ caramel/toffee
- ○ brown sugar
- ○ molasses
- ○ honey
- ○ maple syrup
- ○ chocolate
- ○ cola
- ○ butterscotch
- ○ fruitcake
- ○ gingerbread

FRUITY
- ○ apple
- ○ apricot
- ○ cherry
- ○ currant
- ○ dark fruit
- ○ lemon
- ○ lime
- ○ orange
- ○ peach
- ○ banana

- ○ coconut
- ○ mango
- ○ pineapple

WOODY AND NUTTY
- ○ oak
- ○ cedar
- ○ sandalwood
- ○ coffee
- ○ almond
- ○ walnut

- ○ hazelnut
- ○ pecan

EARTHY AND VEGETAL
- ○ flowers
- ○ grass
- ○ herbs
- ○ tobacco
- ○ tannins
- ○ leather
- ○ salt

- ○ medicine
- ○ metal
- ○ smoke

ALCOHOLIC
- ○ whiskey
- ○ sherry
- ○ brandy
- ○ cognac
- ○ wine

OTHER

STYLE
Choose One

- ○ Neat
- ○ With Water
- ○ On the Rocks
- ○ Cocktail

COLOR
Circle One

TASTING WHEEL
Rate your tasting experience, 1 (lowest) to 5 (highest)

TASTING NOTES

Describe the First Sip

....
....
....

Describe the Third Sip

....
....
....

Describe the Fade

....
....
....

What Is Most Striking About This Rum?

....
....
....

Additional Notes

....
....
....
....

QUALITY RATING

COST RATING

OVERALL RATING

RUM
The Details

Brand Name

Producer | Place of Origin | Price per Glass/Bottle

Distiller | Cask Type | Date

Age | ABV/Proof | Tasting Location

NOSE
Choose All That Apply

SWEET
- ○ caramel/toffee
- ○ brown sugar
- ○ molasses
- ○ honey
- ○ maple syrup
- ○ chocolate
- ○ cola
- ○ butterscotch
- ○ fruitcake
- ○ gingerbread

FRUITY
- ○ apple
- ○ apricot
- ○ cherry
- ○ currant
- ○ dark fruit
- ○ lemon
- ○ lime
- ○ orange
- ○ peach
- ○ banana

- ○ coconut
- ○ mango
- ○ pineapple

WOODY AND NUTTY
- ○ oak
- ○ cedar
- ○ sandalwood
- ○ coffee
- ○ almond
- ○ walnut

- ○ hazelnut
- ○ pecan

EARTHY AND VEGETAL
- ○ flowers
- ○ grass
- ○ herbs
- ○ tobacco
- ○ tannins
- ○ leather
- ○ salt

- ○ medicine
- ○ metal
- ○ smoke

ALCOHOLIC
- ○ whiskey
- ○ sherry
- ○ brandy
- ○ cognac
- ○ wine

OTHER
....

STYLE
Choose One

- ○ Neat
- ○ With Water
- ○ On the Rocks
- ○ Cocktail

COLOR
Circle One

TASTING WHEEL
Rate your tasting experience, 1 (lowest) to 5 (highest)

Describe the First Sip

....
....
....

Describe the Third Sip

....
....
....

Describe the Fade

....
....
....

What Is Most Striking About This Rum?

....
....
....

Additional Notes

....
....
....
....

GUIDED TASTING

Write It Out

QUALITY RATING

☆ ☆ ☆ ☆ ☆

COST RATING

OVERALL RATING

RATE IT

Fill It In

RUM
The Details

Brand Name

Producer   Place of Origin   Price per Glass/Bottle

Distiller   Cask Type   Date

Age   ABV/Proof   Tasting Location

NOSE
Choose All That Apply

SWEET
- ○ caramel/toffee
- ○ brown sugar
- ○ molasses
- ○ honey
- ○ maple syrup
- ○ chocolate
- ○ cola
- ○ butterscotch
- ○ fruitcake
- ○ gingerbread

FRUITY
- ○ apple
- ○ apricot
- ○ cherry
- ○ currant
- ○ dark fruit
- ○ lemon
- ○ lime
- ○ orange
- ○ peach
- ○ banana

- ○ coconut
- ○ mango
- ○ pineapple

WOODY AND NUTTY
- ○ oak
- ○ cedar
- ○ sandalwood
- ○ coffee
- ○ almond
- ○ walnut

- ○ hazelnut
- ○ pecan

EARTHY AND VEGETAL
- ○ flowers
- ○ grass
- ○ herbs
- ○ tobacco
- ○ tannins
- ○ leather
- ○ salt

- ○ medicine
- ○ metal
- ○ smoke

ALCOHOLIC
- ○ whiskey
- ○ sherry
- ○ brandy
- ○ cognac
- ○ wine

OTHER
....

STYLE
Choose One

- ○ Neat
- ○ With Water
- ○ On the Rocks
- ○ Cocktail

COLOR
Circle One

TASTING WHEEL

Rate your tasting experience, 1 (lowest) to 5 (highest)

TASTING NOTES

Describe the First Sip

....

....

....

Describe the Third Sip

....

....

....

Describe the Fade

....

....

....

What Is Most Striking About This Rum?

....

....

....

Additional Notes

....

....

....

....

GUIDED TASTING

Write It Out

QUALITY RATING

COST RATING

OVERALL RATING

RATE IT

Fill It In

ENJOYING RUM

RUM
The Details

Brand Name

Producer   **Place of Origin**   **Price per Glass/Bottle**

Distiller   **Cask Type**   **Date**

Age   **ABV/Proof**   **Tasting Location**

NOSE
Choose All That Apply

SWEET
- ○ caramel/toffee
- ○ brown sugar
- ○ molasses
- ○ honey
- ○ maple syrup
- ○ chocolate
- ○ cola
- ○ butterscotch
- ○ fruitcake
- ○ gingerbread

FRUITY
- ○ apple
- ○ apricot
- ○ cherry
- ○ currant
- ○ dark fruit
- ○ lemon
- ○ lime
- ○ orange
- ○ peach
- ○ banana

- ○ coconut
- ○ mango
- ○ pineapple

WOODY AND NUTTY
- ○ oak
- ○ cedar
- ○ sandalwood
- ○ coffee
- ○ almond
- ○ walnut

- ○ hazelnut
- ○ pecan

EARTHY AND VEGETAL
- ○ flowers
- ○ grass
- ○ herbs
- ○ tobacco
- ○ tannins
- ○ leather
- ○ salt

- ○ medicine
- ○ metal
- ○ smoke

ALCOHOLIC
- ○ whiskey
- ○ sherry
- ○ brandy
- ○ cognac
- ○ wine

OTHER

STYLE
Choose One

- ○ Neat
- ○ With Water
- ○ On the Rocks
- ○ Cocktail

COLOR
Circle One

TASTING WHEEL

Rate your tasting experience, 1 (lowest) to 5 (highest)

Legs · Sweetness · Molasses · Fruity · Woody · Nutty · Earthy · Vegetal · Spicy · Leathery · Smoky · Bite · Heat · Body · Balance · Linger

TASTING NOTES

Describe the First Sip

....

....

....

Describe the Third Sip

....

....

....

Describe the Fade

....

....

....

What Is Most Striking About This Rum?

....

....

....

Additional Notes

....

....

....

....

GUIDED TASTING

Write It Out

QUALITY RATING	COST RATING	OVERALL RATING

RATE IT

Fill It In

RUM
The Details

Brand Name

Producer   **Place of Origin**   **Price per Glass/Bottle**

Distiller   **Cask Type**   **Date**

Age   **ABV/Proof**   **Tasting Location**

NOSE
Choose All That Apply

SWEET
- ○ caramel/toffee
- ○ brown sugar
- ○ molasses
- ○ honey
- ○ maple syrup
- ○ chocolate
- ○ cola
- ○ butterscotch
- ○ fruitcake
- ○ gingerbread

FRUITY
- ○ apple
- ○ apricot
- ○ cherry
- ○ currant
- ○ dark fruit
- ○ lemon
- ○ lime
- ○ orange
- ○ peach
- ○ banana

- ○ coconut
- ○ mango
- ○ pineapple

WOODY AND NUTTY
- ○ oak
- ○ cedar
- ○ sandalwood
- ○ coffee
- ○ almond
- ○ walnut

- ○ hazelnut
- ○ pecan

EARTHY AND VEGETAL
- ○ flowers
- ○ grass
- ○ herbs
- ○ tobacco
- ○ tannins
- ○ leather
- ○ salt

- ○ medicine
- ○ metal
- ○ smoke

ALCOHOLIC
- ○ whiskey
- ○ sherry
- ○ brandy
- ○ cognac
- ○ wine

OTHER

STYLE
Choose One

- ○ Neat
- ○ With Water
- ○ On the Rocks
- ○ Cocktail

COLOR
Circle One

TASTING WHEEL

Rate your tasting experience, 1 (lowest) to 5 (highest)

TASTING NOTES

Describe the First Sip

....

....

....

Describe the Third Sip

....

....

....

Describe the Fade

....

....

....

What Is Most Striking About This Rum?

....

....

....

Additional Notes

....

....

....

....

QUALITY RATING	COST RATING	OVERALL RATING

ENJOYING RUM

RUM
The Details

Brand Name

Producer

Place of Origin

Price per Glass/Bottle

Distiller

Cask Type

Date

Age

ABV/Proof

Tasting Location

NOSE
Choose All That Apply

SWEET
- ○ caramel/toffee
- ○ brown sugar
- ○ molasses
- ○ honey
- ○ maple syrup
- ○ chocolate
- ○ cola
- ○ butterscotch
- ○ fruitcake
- ○ gingerbread

FRUITY
- ○ apple
- ○ apricot
- ○ cherry
- ○ currant
- ○ dark fruit
- ○ lemon
- ○ lime
- ○ orange
- ○ peach
- ○ banana

- ○ coconut
- ○ mango
- ○ pineapple

WOODY AND NUTTY
- ○ oak
- ○ cedar
- ○ sandalwood
- ○ coffee
- ○ almond
- ○ walnut

- ○ hazelnut
- ○ pecan

EARTHY AND VEGETAL
- ○ flowers
- ○ grass
- ○ herbs
- ○ tobacco
- ○ tannins
- ○ leather
- ○ salt

- ○ medicine
- ○ metal
- ○ smoke

ALCOHOLIC
- ○ whiskey
- ○ sherry
- ○ brandy
- ○ cognac
- ○ wine

OTHER

STYLE
Choose One

- ○ Neat
- ○ With Water
- ○ On the Rocks
- ○ Cocktail

TASTING WHEEL

Rate your tasting experience, 1 (lowest) to 5 (highest)

COLOR
Circle One

TASTING NOTES

Describe the First Sip

....

....

....

Describe the Third Sip

....

....

....

Describe the Fade

....

....

....

What Is Most Striking About This Rum?

....

....

....

Additional Notes

....

....

....

....

GUIDED TASTING

Write It Out

QUALITY RATING	COST RATING	OVERALL RATING
☆☆☆☆☆		

RATE IT

Fill It In

RUM

The Details

Brand Name

Producer | Place of Origin | Price per Glass/Bottle

Distiller | Cask Type | Date

Age | ABV/Proof | Tasting Location

NOSE

Choose All That Apply

SWEET
- ○ caramel/toffee
- ○ brown sugar
- ○ molasses
- ○ honey
- ○ maple syrup
- ○ chocolate
- ○ cola
- ○ butterscotch
- ○ fruitcake
- ○ gingerbread

FRUITY
- ○ apple
- ○ apricot
- ○ cherry
- ○ currant
- ○ dark fruit
- ○ lemon
- ○ lime
- ○ orange
- ○ peach
- ○ banana

- ○ coconut
- ○ mango
- ○ pineapple

WOODY AND NUTTY
- ○ oak
- ○ cedar
- ○ sandalwood
- ○ coffee
- ○ almond
- ○ walnut

- ○ hazelnut
- ○ pecan

EARTHY AND VEGETAL
- ○ flowers
- ○ grass
- ○ herbs
- ○ tobacco
- ○ tannins
- ○ leather
- ○ salt

- ○ medicine
- ○ metal
- ○ smoke

ALCOHOLIC
- ○ whiskey
- ○ sherry
- ○ brandy
- ○ cognac
- ○ wine

OTHER

STYLE

Choose One

- ○ Neat
- ○ With Water
- ○ On the Rocks
- ○ Cocktail

COLOR

Circle One

TASTING WHEEL

Rate your tasting experience, 1 (lowest) to 5 (highest)

Legs, Sweetness, Molasses, Fruity, Woody, Nutty, Earthy, Vegetal, Spicy, Leathery, Smoky, Bite, Heat, Body, Balance, Linger

Describe the First Sip

....

....

....

Describe the Third Sip

....

....

....

Describe the Fade

....

....

....

What Is Most Striking About This Rum?

....

....

....

Additional Notes

....

....

....

....

GUIDED TASTING · Write It Out

QUALITY RATING   **COST RATING**   **OVERALL RATING**

RATE IT · Fill It In

ENJOYING RUM

RUM
The Details

Brand Name

Producer

Place of Origin

Price per Glass/Bottle

Distiller

Cask Type

Date

Age

ABV/Proof

Tasting Location

NOSE
Choose All That Apply

SWEET
- caramel/toffee
- brown sugar
- molasses
- honey
- maple syrup
- chocolate
- cola
- butterscotch
- fruitcake
- gingerbread

FRUITY
- apple
- apricot
- cherry
- currant
- dark fruit
- lemon
- lime
- orange
- peach
- banana

- coconut
- mango
- pineapple

WOODY AND NUTTY
- oak
- cedar
- sandalwood
- coffee
- almond
- walnut

- hazelnut
- pecan

EARTHY AND VEGETAL
- flowers
- grass
- herbs
- tobacco
- tannins
- leather
- salt

- medicine
- metal
- smoke

ALCOHOLIC
- whiskey
- sherry
- brandy
- cognac
- wine

OTHER

STYLE
Choose One

- Neat
- With Water
- On the Rocks
- Cocktail

COLOR
Circle One

TASTING WHEEL

Rate your tasting experience, 1 (lowest) to 5 (highest)

Legs, Sweetness, Molasses, Fruity, Woody, Nutty, Earthy, Vegetal, Spicy, Leathery, Smoky, Bite, Heat, Body, Balance, Linger

TASTING NOTES

Describe the First Sip

....

....

....

Describe the Third Sip

....

....

....

Describe the Fade

....

....

....

What Is Most Striking About This Rum?

....

....

....

Additional Notes

....

....

....

....

GUIDED TASTING

Write It Out

QUALITY RATING	COST RATING	OVERALL RATING

RATE IT

Fill It In

RUM
The Details

Brand Name

Producer	**Place of Origin**	**Price per Glass/Bottle**
Distiller	**Cask Type**	**Date**
Age	**ABV/Proof**	**Tasting Location**

NOSE
Choose All That Apply

SWEET
- ○ caramel/toffee
- ○ brown sugar
- ○ molasses
- ○ honey
- ○ maple syrup
- ○ chocolate
- ○ cola
- ○ butterscotch
- ○ fruitcake
- ○ gingerbread

FRUITY
- ○ apple
- ○ apricot
- ○ cherry
- ○ currant
- ○ dark fruit
- ○ lemon
- ○ lime
- ○ orange
- ○ peach
- ○ banana

- ○ coconut
- ○ mango
- ○ pineapple

WOODY AND NUTTY
- ○ oak
- ○ cedar
- ○ sandalwood
- ○ coffee
- ○ almond
- ○ walnut

- ○ hazelnut
- ○ pecan

EARTHY AND VEGETAL
- ○ flowers
- ○ grass
- ○ herbs
- ○ tobacco
- ○ tannins
- ○ leather
- ○ salt

- ○ medicine
- ○ metal
- ○ smoke

ALCOHOLIC
- ○ whiskey
- ○ sherry
- ○ brandy
- ○ cognac
- ○ wine

OTHER

STYLE
Choose One

- ○ Neat
- ○ With Water
- ○ On the Rocks
- ○ Cocktail

COLOR
Circle One

TASTING WHEEL
Rate your tasting experience, 1 (lowest) to 5 (highest)

Legs, Sweetness, Molasses, Fruity, Woody, Nutty, Earthy, Vegetal, Spicy, Leathery, Smoky, Bite, Heat, Body, Balance, Linger

TASTING NOTES

Describe the First Sip

....

....

....

Describe the Third Sip

....

....

....

Describe the Fade

....

....

....

What Is Most Striking About This Rum?

....

....

....

Additional Notes

....

....

....

....

GUIDED TASTING

Write It Out

QUALITY RATING   **COST RATING**   **OVERALL RATING**

RATE IT

Fill It In

INDEX

IMAGE CREDITS